FAUCI'S FICTION

- THE BOOK ON COVID -

DR. MICHAEL J. SCHWARTZ

Published by Michael J Schwartz & Associates, LLC

ISBN-13: 979-8-218-22372-4

Cover design by Marko Markovic
Cover illustration by Michael J Schwartz
Interior Design: Jennie Lyne
Printed in the United States of America

CONTENTS

DEDICATION

This book is dedicated to my father, Joseph R. Schwartz, July 26, 1946, to February 13, 2023. My dad passed away while I was writing this book. I was about halfway through writing when I received the news from his wife. I told him I was in the process of writing this book and wish that he had gotten the chance to read it. This book could have saved his life. My dad was a superhero from a different generation. He was kind, funny, and just plain brilliant. He worked for the Department of Defense for most of his life before retiring in his fifties and had multiple patents on laser targeting systems that he had developed for the US military. He spent a lifetime working in the defense of our country. He was a great man who was loved by his family and lifelong friends. I love you, Dad; you will be missed but never forgotten.

INTRODUCTION

In late 2019, the world started talking about this possible, new, and unknown virus emerging out of China. People were getting sick, and modern technology couldn't detect or treat the afflicted. We were told that this could be the end or that it could be like the Spanish flu of 1918 and that mortality rates could wipe out certain segments of the population. Governments reacted, then overreacted, and the population pleaded for help, guidance, and direction. The virus was identified and given a name: Coronavirus Disease 2019 (COVID-19). Once the illness was synthesized, the laboratories we work with informed us we could begin conducting testing on March 16, 2020. My company was the first to conduct a COVID-19 test in the State of New Jersey.

By this point, the virus was running rampant and spreading quickly from person to person. My company, which consists of two medical facilities and a travel clinic, conducted over forty thousand tests on over nineteen thousand patients. You'd have thought a reasonable person would correlate this test data and use it to calm the masses. However, we had data from as early as March 2020 and thought that something must be wrong considering how the data was being interpreted by the so-called "experts", such as the politicians and talking heads on television. Their responses simply didn't match what we were seeing in practice.

What Dr. Anthony Fauci was saying on television every day during the White House briefings didn't match the actual data we were collecting on the ground. *Did we tell someone? Did we inform our patients about this data?* Well, we tried! Each time we tried to share our data we were looked at like we'd just reported a Big Foot sighting. Nobody wanted to listen because what people were hearing on the news was quite different from what we were trying to disseminate. Our information was 100 percent accurate, but we were unable to share it widely.

I figured that at some point, the masses would catch up and realize what we were seeing in the data. That never happened. Even now, as I write this introduction, there are still groups of people out there who don't understand basic terminology, timelines, or what this virus is and does to those who contract it. More importantly, there are still people who believe in the fallacies that were discussed early in the pandemic. When someone fully understands the facts and data, it looks like most of the world's population was taken for a wild ride, at least with the draconian measures that were instituted. Believe me, Covid is very real, but what you were told about the sickness, how to mitigate it, and what precautions you should take to avoid catching it mimic the makings of a

Hollywood blockbuster. What adds insult to injury is that we *knew* the information being given was misconstrued about a month after testing started! Whether you want to cancel me or not, this is the truth about COVID-19. No BS, no politics, just data.

CHAPTER 1

Perspective

As the first person to conduct Covid-19 testing in my state I have a unique perspective on the pandemic. My data came in very quickly and in mass once my company started working with Covid-19 patients. To start putting all of this into perspective, I want to tell you a little about me, my company and how I attained all this information on Covid-19 before we delve into the science.

Presently, I run three medical clinics, but back in 2019, I only had a traveling clinic. This clinic visited schools, police departments, corporations, etc., to bring medical staff to people on-site. I had a lot of clients lined up for calendar year 2020 before the pandemic hit; however, as news of Covid began circulating, each client called and canceled their appointment dates with us one by one. Our first quarter was shaping up to be a bust, and our company was in serious trouble. I was in disbelief every time an administrator called and told us they were going to "hold off until this Covid thing resolved itself." We were headed for failure, on a fast track to bankruptcy, all because of this virus that nobody knew a thing about.

I am not a medical doctor. My doctorate is in Business Administration; I'm a numbers guy. My background is very diverse, though. I hold three business degrees, I'm a former police officer and private pilot, and I have been an entrepreneur for the better part of thirty years. I own and have run multiple companies over the course of my career, sat on various boards, run a national charity, and consulted with physicians for many years on ancillary laboratory testing, including genetics and respiratory pathogen panels. I hold a few medical certifications as well, but I do not treat patients myself. My staff does.

My staff consists of allopathic and osteopathic medical doctors, as well as advanced practice nurses, medical assistants, office personnel, and part-time clinicians. Our office fits into the category of primary care and integrative medicine with a focus on wellness. As an entrepreneur and business leader, I've also consulted and run my share of political campaigns, so I know how politics works. Politics and medicine are both my passions; however, in my opinion, the two are a bad mix. I think that's why I gravitate toward both of those because they are so different and give me a chance to switch gears. Oh, and by the way, I've also been a traveling comedian for the last twenty years, off and on. So, if I throw a joke in this book here and there, please know I always try to find the funny in serious situations.

Joking aside, politics should stay far away from medicine, and practitioners shouldn't be so heavily influenced by political discourse. Medicine isn't a perfect science; it's a practice requiring trial and error. Science only evolves when people challenge the status quo. Scientists ask questions, test hypotheses, and come up with new solutions through data analysis and interpretation. As I said numerous times throughout the pandemic: medicine is not "one size fits all." No two cancers are the same, nor is every stage of the disease. If all cancer diagnoses had the same treatment, something would be greatly amiss. Yet, this was the approach the powers that be used to handle the entire pandemic. If anyone from a medical background attempted to question their hypotheses, they were shut down, systematically canceled, and shunned by their own medical community.

I have often wondered why people with medical degrees would possibly give advice regarding something they know absolutely zero about. For example, my office is primary care. We know nothing about oncology, and we don't pretend to! When we suspect cancer, we send those patients off to someone with that expertise. Many of my newer patients have told me that their original practitioners shut their doors when the pandemic started. They also expressed their frustrations that their children's pediatricians shut down as they essentially waited out the situation at home. However, all these practitioners who refused to participate during the pandemic seemed to all have "expertise" about the pandemic when questioned on it. Why a pediatrician or any other practitioner who closed their own office when the pandemic started and who didn't test for, treat, or even take a moment to consult with someone who understands virology would recommend anything related to COVID-19 is beyond me. These practitioners lacked the knowledge and perspective about COVID-19, yet the public relied on them heavily for advice. It's no wonder why the public lost faith in the medical community and the CDC.

When someone is telling a story, it's important to understand their perspective. Perspective is crucial for the reader to understand how information may be interpreted if there happens to be any bias, and to determine whether the information itself is credible. When someone gives their opinion, I inevitably ask the question "based on what?" This becomes very important, and we will touch on this more later. You would think that with an ongoing, possibly deadly, worldwide pandemic, everyone would want the facts and nothing else. It still amazes me that even today, three years into the pandemic, so many people lack an understanding of the language related to the virus, and that includes practitioners and medical personnel. Language and terminology are important because if we all say one thing but think another, nobody makes any progress, and we simply stay in the status quo. It frustrates me everyday that when I say something to a patient, they will nod in agreement but later ask me a question where I realize they didn't know what I was talking about in the first place. I've completed extensive research on COVID-19, and my perspective is a little different from that of most people I encounter. Or at least, it was very different in the beginning. The masses seem to be catching up slowly as they have more time to digest the data for themselves but it's too little, too late!

In January 2020, the media cited stories from across the globe about the pandemic. The majority of Europe was in lockdown, and death counts were rising. It seemed like most of the United States was watching from afar, thinking that this was a problem for China and Europe but that nothing serious could happen here. Thus, the blame game started and centered on China. Was this a zoonotic leap from animals to humans, or was this virus synthesized in a laboratory and somehow released accidentally (or purposefully)?

I was collecting N-95 masks, shopping for non-perishables for my pantry, and sounding like an alarmist to my friends before anyone in the United States was even talking about this. My friend is an ICU trauma nurse at a local hospital in New Jersey, and, quite frankly, she thought I was losing my mind. I went food shopping and filled the cart on a Monday, and by Wednesday, I told her I was headed out to the store again. She laughed and thought I was overreacting. Not one person on the news or in government was considering Covid with any seriousness. When I went to the store again on Friday to load up another cart, my friend had serious concerns about my well-being, as this was weeks before the panic-buying started. However, I was simply following the trend out of Europe, as I could see how this could refashion over to the United States very quickly.

To add some additional perspective, back In January 2020, I came down with what seemed to be the worst flu I'd ever had. I distinctly remember not being able to taste anything, and my smell was off. This was well before anyone discussed the specific symptomology of Covid-19, like the sudden loss of taste and smell. This "flu" lasted for weeks; the cough was persistent, dry, and seemed to come on at the worst possible times. The fever would occasionally subside but would run higher at night, which is normal for an illness, and I would wake up with my bed sheets absolutely soaked. I was starting to wonder when this sickness I had would finally break for good.

I was consulting on a congressional campaign as a senior advisor to my candidate David Richter at the time, who was running for a seat in New Jersey's Third Congressional District. We went to a presidential political rally on January 28, 2020, and I thought to myself, *I must be reaching the tail end of this soon.* I wasn't thinking about trying to protect others because I thought I was getting over the flu, and we all just got on with it when we were sick back then. If I was able to get myself out of bed, I would go to work! I had stuff to do, and I wasn't going to let a cold, flu, or whatever keep me from my obligations.

David and I had about an hour's drive to reach the rally. There we were, breathing the same air, prepping for his big evening and knowing that just minutes after we exited the car, he would be backstage shaking hands with the president of the United States and all the dignitaries in attendance. After this, he would take the stage himself to give a speech and introduce the president to a crowd of thousands. I remember parking in what was essentially no man's land, as the police and secret service had very tight access to the venue. The walk to the auditorium was horrible; the evening temperatures were in the midtwenties. We were at the beach, and those Jersey Shore winds can be brutal during the winter. I couldn't shake off the chill going through my body.

On the walk in, I spotted former Saturday Night Live alum Joe Piscopo. He was in line waiting for entrance, and we stopped to say hello as we were being ushered up to the front by security. I had worked with Joe a few times through my charity, and he was always a gracious guy. I introduced David to Joe, and David was surprised that Joe knew exactly who he was! I'm guessing Joe saw some of our political commercials, but he always had his finger on the pulse of local, regional, and national politics. I felt great because it gave David a little boost before going in to give his big speech.

When we got inside and eventually backstage, I could barely stand up straight. Whatever I had going on wasn't just a little cold. This thing I had assumed was the flu was knocking me on my ass. I somehow made it through the event, but not one word of this virus that was running through China and Europe was even mentioned. I couldn't wait to get home to my bed and just work whatever was lingering in my system out of me.

Two short weeks later, on February 11, 2020, the World Health Organization (WHO) gave the mystery virus its official name: COVID-19, which stands for "Corona Virus Infections Disease 2019". About two weeks later, and with my cough still lingering, I was back out with my candidate at another political function. I had been watching what was going on in Europe closely, with a keen emphasis on Italy. David, our Campaign Manager Tom, and I were standing next to the bar in the meeting hall, waiting for the event to start.

Out of earshot of the crowd, I said to them both, "This Covid thing is going to explode!"

They both looked at me, smirking, before David said, "Come on. This is going to be gone in a week."

I said, "I don't think so. Look at what's going on in Europe—"

David cut me off and told me, "I survived H1N1, Ebola, and about a million other things. I think I'll survive this one, too."

I could tell I was losing the room here, but I kept at it. I said, "I'm not worried about the virus; I'm worried about the government's reaction to it. This thing is going to be a total disaster." Even today, that conversation sticks out in my head as Covid completely changed the way that year's election was run. The intermixing of politics and healthcare had begun, and it wasn't going to be pretty.

My perspective on COVID-19 started on March 12, 2020, just four weeks after that political meeting and two short months after the Presidential political rally. This was around the time that the virus had finally been synthesized. Labs were now able to test for Sars-Cov-2, which is the technical laboratory term for Covid-19, but I still found myself pondering the future of my companies. Everything seemed bleak, and as the world was being systematically shut down, I wondered to myself how I would economically survive this pandemic. After weighing my options, I decided to take a flight from New Jersey to Fort Myers, where some good friends live. My dad and my stepmother lived a few blocks away from them too, and I thought this might be a good time to kill two birds with one stone. I needed to clear my head, and I do my best thinking when I leave my own bubble.

When I arrived in Fort Myers, my friends were obviously concerned about this new virus. They are a bit older, retired, and did very well for themselves in life. Years ago, they sold their company for a hefty profit and have since become philanthropists. Most of their lives were now spent with family, traveling, and attending charity events. Their concerns weren't the same as mine. While I was concerned about my future and how I was going to survive without maxing out my credit cards, they were solely concerned about their health and the prospect of possibly dying if they encountered this thing. Both friends had some serious health concerns, including diabetes and a history of lung cancer, so I was also concerned about their well-being and the prospect that they wouldn't be able to weather this thing as easily as I could.

They have a beautiful vacation home in Fort Myers, Florida, with a very comfortable screened-in back deck with a wonderful lake view. It's a great place to sit and forget that the rest of the world exists. Phil and his wife Marilyn are smokers, so on every trip that I've ever made to their place, we immediately head to the back so Phil can smoke, sit outside, and chill out until we head to dinner. I sat there with Phil and started showing him some regression analysis charts that I'd run before coming down. I took some early data from the COVID-19 infection numbers and ran some predictive analysis. At the time, most countries were reporting under one thousand cases, and most still had under one hundred in total.

If you remember, the United States only reported the nation's first case of Covid-19 on January 18, 2020, in Washington State and at that time Covid-19 didn't even have its name yet. I didn't have a lot of data to work with considering we had only been tracking it for approximately 2 months. Everyone was referring to a "novel virus out of China". My trip to Fort Myers was during this time when most people were completely oblivious as to what could develop. I think most people, including Phil, thought this thing was a problem for China and maybe Europe, but it would never really take hold in the United States. Phil wasn't overly concerned.

I took a green Post-it pad that was sitting on the table and wrote down some numbers. I said, "Phil, you need to take this thing seriously and think about flying back to New Jersey this week." If they needed to get home quickly, I knew that it could become an impossibility once Covid got out of hand. I showed Phil what my predictions were for infections in the United States one week out, then two, three, and four. Next, I showed him what I thought the infection rate would be in two to six months from that day. The numbers were absolutely staggering, and Phil's eyes opened wide once he saw the pad. Nevertheless, he remarked, "Come on, really?" as though he thought my numbers seemed too farfetched.

I pleaded with him a few times over the course of my trip to come back to New Jersey since that's where their main residence, their hospital, their doctors are, and their close friends reside. With their serious health concerns, my fear was that if travel, supply chains, and life in general were interrupted, they may end up stuck at their vacation home, which could cause more problems in the long run. Once we got through the welcomes and the eight-hundred-pound gorilla in the room, we moved on to dinner and tried to relax. The following day, my conversation centered on what I was going to do with my life, considering all my clients had called to cancel the previous week. I was more than concerned that I was heading for bankruptcy.

On March 13, while still in Florida, my phone rang. It was a call from one of the boutique labs we work with. My contact at the lab wanted to gauge my interest in COVID-19. She asked, "Are you getting any interest from your clients about testing for Covid?" She told me that they had applied for and received FDA approval for Covid testing and that they would be legally accepting samples as of next week on March 16.

Phil heard the entire conversation, and as soon as I hung up, he said, "Well, what do you think?"

I paused for a moment, took a sip of my drink, and said, "Well, I've got nothing else to do, so I'll have to take a hard look at it." I sat there for about all of thirty seconds before immediately calling my staff back home and informing them of my intentions. My assistant asked how I wanted to do this, and I responded, "I have no clue, but I'll figure it out somehow like I always do."

When I flew back to New Jersey the next day, I remember the plane wasn't even halfway full, and there was this eerie silence on board. I distinctly remember thinking to myself that this might be the last flight I would be on for a while, and I couldn't help but recall a show I'd watched on the History Channel a few years back about how the world would react during a pandemic. Ironically, the world was reacting exactly as experts had predicted they would. Bravo, History Channel. I think the show was called *After Armageddon*. I highly recommend it if you get a chance. I couldn't get that show out of my head while I was watching the reactions of the governments around the world. Once Europe started to freak out, and I couldn't help but think this was all foreshadowing where we would be in the United States sooner than later.

CHAPTER 2

The First Day of Testing

I immediately went to my office and started taking an inventory of every lab sample I had after landing back home in New Jersey. Our partner lab had been conducting respiratory pathogen panels (RPPs) for many years, and we had lots of samples and swabs lying around. An RPP is something that tests for things such as colds and flu, which consist of both viral and bacterial pathogens. In a lot of instances, these RPP's are not deemed medically necessary by the insurance companies so they won't be covered. Part of the problem with modern medicine is that, in most cases, insurance dictates care and influences physicians on how they practice. Practitioners are always trying to balance what a patient needs versus what their insurance company may send them a bill for. These instances can cause a larger problem especially when it comes to over prescribing and causing antibiotic resistance. Forty-one million antibiotics are prescribed every year, of which twenty-three million are completely unnecessary. Patients today are so used to a world of instant gratification that they tend to get upset if they visit a physician's office and leave empty-handed. While it is predominantly patients driving this narrative, physicians are also to blame as they sometimes hand out scripts far too quickly to please patients.

The problem with handing these prescriptions out so freely is that, over time, people develop antibiotic resistance. Sometimes, patients self-diagnose and will demand antibiotics or take old medications that are lying around, which aren't the appropriate course, which speeds up antibiotic resistance. Once people develop resistance, there's less chance those same antibiotics will work as intended when they're needed the most. Although our practitioners will still write a script for a patient based on symptoms alone, our protocol is to swab a sick patient and send it to the laboratory to conduct an RPP so we know what we are treating. It helps to know precisely what the patient has because viral pathogens don't respond to antibiotic treatment, but bacterial pathogens will. Then, once the RPP results come back, our practitioners will take the patient off the prescribed antibiotic if necessary.

That's why our own patients knew what Covid swabs were before they realized that they were "Covid" swabs! We had used swabs for years and had pleaded with other physicians to implement RPPs in their offices but were always met with resistance. Once, a doctor told me, "I've only got seven minutes with a patient, and now you want me to pull out a swab and fill out a form?" Others would say, "I've been writing meds for thirty years. Now I have to look at a report to figure out what I need to write?"

My team and I have been using state-of-the-art testing for years, including pharmacogenetic testing and cancer genetics, etc. Most physicians don't receive more than one day of training on things like pharmacogenetics in medical school; nonetheless experience it before opening a primary care facility. Physician's God complexes tend to obstruct good practice and actual patient care in many cases. Most primary care offices are concerned mostly with volume, because quite frankly, there is no money in primary care, and offices need to churn these patients in and out as quickly as possible just to cover their overhead.

RPPs were second nature in our office, so if our lab partner was getting approval to add COVID-19 to their panel, this would give us a unique opportunity to keep our company afloat and to be first to market! I gathered up all the current RPP supplies and waited patiently for the lab to send over the COVID-19 requisition form. I called a few friends to let them know what our company's intentions were, as well as one of our municipal partners who we had completed some wellness work for in the past. On Monday morning, March 16, 2020, the lab promptly sent over an electronic requisition form. Twenty minutes later, I received a call from the mayor of our municipal partner, asking for help.

Our county's first Covid case was in the mayor's township. The patient was hospitalized and on a ventilator. The patient's prognosis was grim, and his family had been in touch with the township, requesting to be tested for

Covid. Luckily for me, and I guess for them, too, they had called the right place at the right time, and eventually, that call filtered down to me. There were four family members, and the medical testing regulations at that time caused a serious problem. They were part of the same household and had been exposed to the patient; however, only two of them were experiencing any relevant symptoms. The regulations clearly stated that a full RPP could only be conducted out of medical necessity to run a full respiratory pathogen panel, and the lab hadn't sufficiently separated COVID-19 as singular item apart from running a full panel of tests.

You see, a practitioner can use a test to check for one specific ailment or can just check off the entire panel. There is a method to this, so bear with me. Panels are expensive, and as I mentioned, insurance dictates healthcare, so medical necessity comes into play here. If you went into a doctor's office for flu but received a negative test, would the doctor just suspect a common cold and treat it as such, or would they run additional tests to identify what the pathogen is? When someone is sick, our RPPs can detect thirty-one potential sicknesses. Most of the time, Human Coronavirus (a common cold) isn't the culprit. When I visited the family, I was only legally allowed to run full panels on the two people who were sick, and, unfortunately, I was only permitted to test these two sick individuals for Covid, too.

I was only allowed to test two of the four individuals in the house, but I knew the odds were that if they were exposed, it would benefit the family to know who had what. When I got to their house, I was shaking nervously.

I've done a lot in my life, so nothing really fazes me, but when I reached the family's house, I had to think very seriously about the approach I would take and how I would handle testing them. I called the family from the car and asked if they would come outside onto their porch, one at a time. When they agreed, I donned some personal protective gear (PPE) and walked toward the house.

One by one, they came out onto their porch as I juggled swabs, sample containers, and requisition forms. This wasn't normal practice to go to someone's home and make house calls, especially during a national emergency, but this was what was necessary. Cognizant of their sick family member's health situation, I asked how he was doing. They were concerned but as optimistic as they could be considering the bad news of Covid spreading around the globe. I couldn't help but feel powerless; this poor family was going through such a terrible time. However, I was also extremely careful of what I touched and how I touched it, and well aware that this invisible disease could potentially get me sick too.

After testing everyone, I walked back to my car. I had left the trunk open beforehand and carefully grabbed some disinfectant towelettes. I proceeded to wipe down anything and everything. I carefully removed my surgical mask and put it into a disposable bag. I then took off my N-95 and any remaining PPE before bagging the samples and heading over to the local FedEx to send them off to the lab. I sent a text to the mayor to let him know that the job was done, and I patiently awaited the lab results.

I work with a few boutique labs that aren't large monstrosities like LabCorp and Quest, and are realistically the only "in-network" labs in the United States. Boutique labs are great because they run extremely beneficial ancillary tests, and their turnaround times are lightning fast as compared to those "in-network" behemoths. You can always get a representative on the phone to discuss issues, and patient care is done in real-time. Even though boutique labs are "out-of-network," the patients get much better value from them. See, when an out-of-network lab has a balance that the insurance company doesn't cover, they try to collect because they legally must, but they never put patients into collections. Therefore, if a balance is unpaid by the patient's insurance, it will never affect the patient's credit. When one of those behemoths is short even thirty bucks, they will come after you, and if you don't pay . . . well, good luck trying to get all those past-due balances off your credit report.

The following evening, the lab called, asking to speak to our practitioner immediately. Our nurse practitioner was extremely busy between working for us, managing time at a hospital, and handling respiratory-compromised patients. I was able to patch her into a call, but the lab quickly cut me off and asked if they could speak with her privately. Perplexed, I had them both trade numbers and quickly got off the call. About two minutes later, my practitioner called me to tell me what had transpired. The lab rep had said to her in a quiet, somber voice, "You have positive patients." My nurse practitioner thanked them, took the patients' names, and got off the call. Then, she called me back and said, "If this is how they need to handle every positive patient, we're going to have a problem!"

We didn't have time to stop what we were doing every time a patient tested positive, and, quite frankly, it was annoying to have to pull her away from her busy schedule. There had to be a better way. Our partner lab was a tad over-the-top in the early days and could have benefited from a reality check. If they thought we were going to stop everything we were doing to take thousands of phone calls notifying us of results, we were going to have to find another lab . . . and quick. Even today, that call reminds me of what life was like before Covid, when privacy and HIPAA mattered. What a different world we live in now!

After the call, my team and I came up with our own treatment protocol that was simple and made sense: rest, drink plenty of fluids, take vitamins C, D, and Zinc, and if patients experience any respiratory difficulties, they were to go to the hospital. They would also be asked to isolate for ten days until their symptoms had subsided before going out in public. My nurse practitioner asked me if I could call the positive patients and relay that protocol word for word every time. So, that's what we did, one positive test at a time. When we got the call from the lab that night, in regards to that first family we tested, it set off a chain of events that would dictate the next two years of my life... I took on the responsibility of calling every single positive patient to go over instructions and record their symptoms and, in most cases, talk them off a ledge.

When the results were ready that evening, I called our first family of patients. All four of them were positive, but only two had any symptoms. Neither of those who had symptoms needed any medical attention, as they were mild symptoms at best. I inquired about the hospitalized patient, their father, and was told he was in the same boat as the day before. The family were optimistic but were losing hope quickly. I told the family the protocol and recommended Tylenol for any aches, pains, or fever. I also told them to avoid Ibuprofen, as it was thought at the time that this would cause Covid to flare up.

A few days later, I called to check in on that family. All four of the positives were doing fine. The two with symptoms were on the upswing, and the two without symptoms never developed any. However, the entire family was overtaken by grief; their father had passed in the hospital as he succumbed to the effects of COVID-19. He was the first recorded COVID-19 death in Ocean County, New Jersey.

You will see, as we saw back then, that there is a lot to digest as we move forward. We were trying to figure this all out in real-time and weren't getting much help from the federal, state, or local governments. I think it's important to give you my perspective in a timeframe to properly understand our frustrations with the Covid pandemic and see how the data correlated to each period. It's also imperative that you understand when and why we knew very salient things well before the public was ever informed by the media. I will take you through our experience in the order it happened and will wrap all this up with the facts because the facts won't lie.

CHAPTER 3

The Language of a Pandemic

I realized early during this crisis that language, phrasing, and verbiage are very important. The world had just been thrown into a global pandemic, and everyone was on a completely different page. The media pundits were going in twenty different directions, and most people had a lot of questions. *Do masks work? What about a vaccine? What if I'm elderly and have comorbidities?* This confusion was probably the main catalyst for my writing this book. I was writing this three years into the pandemic, and I still had patients coming to my office who knew nothing about how viruses transmit. They knew nothing about incubation timelines, what to do in the event of a positive diagnosis, or about the hundreds of other Covid-related things we get questioned on daily.

I didn't realize how important language was until one day, in March 2020, I was schooled by an epidemiologist during a call. We had been discussing a positive patient, and I had told the epidemiologist that we recommended they quarantine for ten days. He stopped me in my tracks and said, "What? You don't quarantine when you are positive; you isolate. There IS a difference!" This one interaction in the infancy of the pandemic made me realize that if the population didn't get on the same page quickly, the world would be in trouble. How would we get a handle on Covid?

The lab language alone can be quite confusing. A lot of people would just call this the Coronavirus. Some would say Covid, while others would say Covid-19 and the lab would just reference Sars-CoV-2. *What is Sars-CoV-2?* you ask. Medicine can be extremely confusing to the layperson, including those of us whose degrees are in other fields. However, I often find myself explaining the lab portion of what we do to physicians who have zero clue as to how to read a report or put the relevant information into practice. Severe Acute Respiratory Syndrome Coronavirus 2, or Sars-CoV-2 (lab speak for Covid-19), would appear on requisition forms, and physicians would call me and ask if they were testing for the right thing: "*You sent me the wrong requisitions form!*"

A reasonable person would think that if an expert came on television, maybe a month or so after the pandemic really took hold, and spoke to a national audience, giving a twenty-minute dissertation on timelines, illness, and precautions, we would all be on the same page. The sooner we could all speak the same language and effectively communicate about COVID-19 intelligently, we could all move on and get back to our lives. However, the "experts" on television were all saying different things. Most of them weren't "experts" at all. They were all just happy hearing themselves talk and kept reiterating what other "experts" said without having any real data to back it up.

Even the "experts" at the White House were contradicting themselves daily, and nobody dared check in with the practitioners who were dealing with this on the ground to collect real data. It generally took anywhere from six to twelve months before the talking heads on television caught up to what we were saying to our own patients in our offices daily. This is still true today from testing, masking, vaccination, and treatment. We could gauge the early data, see trends, and make reasonable predictions. This is how science works, yet all the talking heads were screaming as loudly as they could that they "follow the science."

People go to the doctor when they're sick; however, countless doctors' offices shut down as they were scared and didn't have a clue how to tackle this thing. Doctors were much more concerned about bringing something back home to their own families than they were with treating patients. In other words: doctors' offices all closed just when the world needed them the most. We had patients showing up at my offices asking to be tested for Covid so that they could see their regular doctor. Yes, you needed to make an appointment at a different primary care office than your usual one so you could get a Covid test to go and see your regular doctor. Nevertheless, this was great marketing for us since we converted most of these patients to our own practice after they realized the irony of the situation.

I was hoping these other practitioners would participate in the process so all of us could share data and figure out the nuances of Covid a little quicker. However, we were essentially left on an island, alone to decipher data and to make sure the language of Covid was succinct. So, let's make sure we're on the same page and speak the same language. I never thought we would be here at this juncture, but still, three years into this (as of the time of writing), I have patients who I call to let them know they are positive, and they ask me if they need to "quarantine." To which I respond "*No*, what did the astronauts do when they came back from the moon? They quarantined in case they caught something. You are positive, so you need to *isolate*. There is a difference!" Hats off to that epidemiologist!

The incubation period for COVID-19 is three to five days, and the virus can only sustain in your system for about fourteen days. That incubation period is consistent with a lot of other respiratory pathogens. So, by the time you started feeling symptoms and walked into one of our offices, you would have had Covid for about four days. Once we notified you that you were indeed positive, you would need to stay home and isolate for ten days. Four plus ten equals fourteen days which is the lifecycle of the virus rendering you no longer contagious if you present with no symptoms and are back to your normal baseline health, which put you at fourteen days, where you are no longer contagious.

We disseminated the above information for the entirety of the pandemic. However, as the pandemic dragged on, the CDC changed the isolation period to five days once they realized that people had been out of work for so long, which had affected supply chains. Or at least, that's what was reported in the news. Despite the "science" remaining the same, it seemed as though this disruption in essential services was the only reason the CDC would make such a dramatic change in the Covid guidance. Don't get me wrong, science is constantly evolving as new information is tested and challenged, but the timing of the CDC's decision was a little ironic. The bottom line is: if you still have moderate symptoms, you're still contagious up until the fourteen-day mark. This was one of the points where it didn't seem like the government was really bothered about "patient care." Although, if you read one of the original World Health Organization reports on Covid spread, for anyone who presents as asymptomatic the spread is virtually zero. So then, why did governments have every human on the planet running around in cloth masks? We will visit the mask topic a little later.

One of the other issues with language is the use of the term "coronavirus." There are many identifiable coronaviruses floating around, and we test for seven of them in our office.

When a patient comes in and tells me that they have "Corona," I ask them whether they have a beer.

When they say, "No, the *virus*," I ask them whether they have a common cold. The common cold is also a coronavirus.

When they say "COVID-19," we then start to narrow it down. Not to sound like a jerk, but we were dealing with a global pandemic, and nobody was taking this seriously.

How were we ever going to get out from under it? Most people have never taken a class in virology and wouldn't know the first thing about transmittal, protection, or the therapies used for them. Therefore, I think education is the first step. When a patient starts asking questions about Covid, I usually tell them what a virus "actually" is, first and foremost. Once we get that and a twenty-minute speech on testing, masks, and just about anything else Covid-related out of the way, patients usually look at me and say, "You should write a book."

I guess I heard that one too many times and finally decided to put the figurative pen to paper.

However, once you start to see how the actual science that we observed conflicts with the "science" the government told you existed, you may be a little upset that I didn't write this book a bit sooner. We knew most of what I'm telling you right now relatively early on during the pandemic. We knew it while the lockdowns were still fresh, and we could have minimized the disruptions to our supply chains that followed. We could have minimized the damage the pandemic did to our nation's children being masked up and out of school. Why did the government ignore our data? We can all speculate as to why. My purpose and intent aren't to lead you down a certain path of political narrative; I will simply report the facts and when we learned them. You can make your own decision as to the who, what, where, when, and why. The bottom line is that you were lied to on a massive scale, and most of the population bought that lie, hook, line, and sinker!

CHAPTER 4

The Chaos Has Started

After I conducted those first tests in March 2020, my phone started ringing off the hook. Word of mouth spread quickly, and just like the run on toilet paper in the grocery stores, I was bombarded with calls for people clamoring for a test. Imagine that! People were falling over each other just to get a swab stuck up their noses. Nowadays, you couldn't even pay someone to get swabbed, but when this all started, those swabs were worth their weight in gold.

The lab was telling me that there was a worldwide shortage of swabs and medium, the fluid that the swabs go into in the tube. This shortage became a major-league problem that lasted for quite a few weeks. We couldn't conduct tests because there was no fluid to keep them stable on their way to the lab, so we had to turn a lot of business away during that time. The supplies we had in stock each week were generally used by Tuesday at the latest. If the lab couldn't resupply us, we would have to close from Wednesday through to Sunday.

Before the pandemic, I could just call the laboratories I work with to send us any testing supplies we were running low on, and we would receive them within two days. Once the pandemic began, this all changed. This was uncharted territory. Our partner labs started calling me daily to let me know they were having trouble sourcing supplies. To be completely honest, these labs were pros, and I figured they would work out the problem and, in a few days, we would be back on track. Boy, was I in for a rude awakening. I had enough inventory to get started with Covid testing, but as the calls kept coming in, I had to frantically call those labs and search the open market for whatever I could find. I started running out of testing supplies daily. I scoured the internet for anything I could find, and each time I thought I had hit the jackpot, the order confirmation quickly turned into an apology email stating that my order couldn't be shipped.

I then started going back through my records and called every single lab I had ever worked with over the years. All of them were having trouble sourcing everything from swabs and medium to the chemicals and plates used in PCR testing. There was a worldwide shortage of everything and anything, and this is when my fears started to become a reality. I had just transitioned out of near bankruptcy, and here I was again, faced with the real possibility that my company would be out of business in just a few short days.

Remember the great toilet paper shortage of 2020? It sounds funny when you think about it now. I had already stocked up if you can recall my own personal lead-up into this pandemic. In fact, I didn't need toilet paper, paper towels, or soup during the entire pandemic. I still have plenty of soup. If anyone wants to come over for dinner, I'd be happy to host you. The entire situation was even more surreal in the weeks that followed, seeing entire isles of products just wiped out in every store you could get into. I remember one day, I just needed some pasta sauce for dinner. I'm mostly Italian, although you wouldn't know it from my last name. Although, I do know I'm going to get in trouble for not making my own sauce or, as some say, "gravy" once my family reads this. Nonetheless, the things we were all used to having easy access to were now hot commodities. I never thought I would see grown adults fight over the last roll of toilet paper in a supermarket, but it happened!

While we were the first and probably the only company in New Jersey conducting any kind of Covid testing for the public, there were hospitals running tests, but they were sending them out to those large, unreliable labs, which took seven days and, in most instances, just ran Covid-only tests, which we will get into later. So, if you needed a test and wanted it fast, I was the guy to call! People were starting to freak out, and I didn't want to get to be treated like the last Cabbage Patch doll on Black Friday. Some people were even comparing this to Middle East Respiratory Syndrome (MERS), which had a 30 percent death rate, and those same people were preparing for the end of times.

When we conducted that first test on March 16, 2020, we couldn't imagine how quickly things would progress but the following day, March 17, the world was effectively shut down. Do you remember the phrase "Fifteen days to stop the spread"? I put a reminder in my calendar that said "Closed the World" on that date, just so I could look back and see how long this entire exercise would take. I started getting requests by the hour, and word of mouth had spread quite quickly. Somebody would call me from a random town or area, and over the course of the next few days, I would get even more calls from the exact same location. *"Hey, I got your number from my friend down the street. Can you come over and test five of us?"*

Our company was like everyone's best-kept secret. If my patients knew that we were running low on supplies, they would protect the stash like it was their own. The funny thing is, while nowadays everyone in the world hates getting their nose swabbed, back then, EVERYBODY, and I mean EVERYBODY, wanted a Covid test. People were begging to get the "Q Tip" shoved up to their brains...

I got tired very quickly of patients calling it a Q Tip and telling me I was going to hit their brains. For the record, those "Q Tips" are nasopharyngeal swabs that go into the nasopharyngeal cavity, which runs above the oropharyngeal cavity. Bottom line is the virus is most active in the nasopharyngeal cavity and you'll get your best sample by slightly entering that cavity, making you feel like you're going to sneeze. Medical professionals had already been using these nasal swabs for years when someone presented with symptoms consistent with a respiratory pathogen. It was only with the emergence of Covid that this form of testing became mainstream. Also, yes, I saw what the Chinese were doing and swabbing the "other" cavity. We are just not that kind of practice; sorry to disappoint you. I did have patients ask, though!

In my home state of New Jersey, it was as if the world had stopped in its tracks. I think some people thought that we would truly get a handle on Covid if we just gave the government the fifteen days they asked for. The order from the Governor was for all non-essential personnel to stay home and not leave unless it was essential that you travel, for example, if they needed to go to a doctor's office or the grocery store. The police were also given orders not to conduct traffic stops unless it was necessary, which worked out well for me as I was making house calls and driving up and down the highway at an unbelievable pace.

It was great; there was no traffic on the road, and I was visiting five or six houses a day, most being miles and miles apart with multiple patients per household. I had a first come, first serve policy at the time and took the patients as they called. They couldn't leave their own homes, but I was out and about all day long. I met tons of great people. Most wanted someone different to talk to those first couple of days after being cooped up in their homes. I'm sure the Covid tester wasn't always a pleasant sight, but once they realized I was human too, the need for human interaction took over.

I felt like a character from the movie *Red Dawn*. It was like I was on the front lines every day, shuttling information back and forth to my troops, who were hunkered down, just waiting for the OK to move out. I remember thinking to myself how surreal this seemed. The more people would call me and ask how long these lockdowns would last, I was positive it couldn't drag on much longer, especially once we started collecting meta data. We were SURE the media and government would catch up quickly and say, "OK, let's bring this down a notch." Today as you read this, nobody really cares too much about Covid-19, but back in the first few weeks of lockdown, most people would say the word Covid with a low hush. It was still taboo to speak of, and when anyone died or got sick, the immediate assumption was that it must be Covid.

The only thing that was keeping me from taking on more patients at the time, was our lack of supplies. I had daily calls with our partner labs to go over results, supplies, sourcing, and just about anything else Covid-related. They also finally stopped calling the practitioner with every positive test they received because everybody, and I mean everybody, was coming back positive. Of course, we had our share of negatives, but the frenzy pushed people to test, even if they were only exposed for a second or two. I heard every reason under the sun for people to want to be tested. I was driving long distances to conduct tests for people who had fifteen seconds of contact with someone who'd had ten seconds of contact with a positive case.

We do a lot of work with the local hospital system here in New Jersey, and I got an email one day from the regional president, who asked me how I was getting my results in twenty-four hours while the waiting time elsewhere was up to seven days in most instances. I explained that I had been using boutique laboratories for many years and that I had great relationships with them.

I emphasized that our relationship with them mattered in this instance because even though their turnaround time was normally a day, other entities were trying to send them additional business, which they had to turn down to accommodate their long-standing clients.

When I say "other entities," I mean that state governments, municipal governments, and every Tom, Dick, and Harry was trying to send Covid samples to accredited labs that could process them. This went for other practitioners, too. Most of the ones I know of had zero relationships with outside labs because, quite frankly, they never cared about ancillary testing in the first place. There are some great primary care doctors out there, but there is a new way of treating patients. I think, by now, the pandemic has really demonstrated some of the shortcomings of the medical community.

My team and I had an additional advantage because swabbing was our daily norm even before Covid existed. We were used to swabbing, sending the sample overnight via a courier, and receiving the report letting us know which pathogen the patient had by noon the following day. Covid, for us, was just another pathogen on our rec form. Seemingly, for everyone else, the whole process was a giant clusterfuck. Companies scrambled to get into the testing game, and physicians' offices needed to locate labs able to run testing while educating themselves on how to conduct tests. If these practitioners had simply practiced good medicine in the first place, they might not have had such a hard time playing catch up.

I was able to strike a deal with one of our local hospitals to do some of their testing for rule-outs, which is when hospitals need to rule out COVID-19 as a diagnosis before moving the patient into another area or discharging them. The usual seven-day wait time for testing was impossible for hospitals to work with, as they couldn't hold patients in the hospital for this long without having a clue as to what they were dealing with. Medical facilities were even short on masks and PPE, so it was incredibly risky to keep patients for seven days when their Covid-status was unknown. Most of the hospital staff had to reuse their own N-95 masks for days. The assistant business administrator of one of our municipal partners called me one day requesting I find some masks just so they could transport prisoners back and forth in their police patrol cars.

Luckily, I was able to source some masks from my buddy Mike at a dental supply company for them. I was able to get two boxes. Just two. This was the only order I was permitted to place. Mike told me to point blank that this was all he could do, and it was the last time he could do it. You would have thought we were at war with a foreign country, and he was giving me one last solid in life with how gravely he said it. I paid an exorbitant amount of money for those two boxes of masks, but I was extremely happy to have them.

At one point, I had to ask the hospital for some supplies and was elated when I was able to get a whopping forty swabs in one day. We were only getting about ten or fifteen swabs at a time from the lab sporadically every few days or so. One day, our partner lab had some extra, and I was able to secure twenty-five at a time when I considered anything over fifteen a home run. I remember saying to somebody one day that my goal was to do at least a thousand Covid tests, but at the rate we were going, we would be lucky if we hit a hundred. It was very disheartening as the requests kept coming in, and I had to keep turning business away, which is not in my nature. However, we were stuck between a rock and a hard place.

One day, I got a call from an assisted living center about an hour's drive from me. They were interested in testing their entire population. I was trying to figure out how to accommodate them when I finally got some much-needed good news from the lab. They had come up with something ingenious! They had been able to validate buccal swabs with the FDA. Buccal swabs are usually used for genetic testing, they are much larger than nasopharyngeal swabs, but they worked! In addition, the lab was also able to validate a different liquid medium to go into the falcon tubes, and, in an instant, as if a light switch had been flipped, we were back in business!

Sounds too easy, right? It's amazing how the free market works. These labs were just as vulnerable to going out of business as I was. They had the expertise and drive and just kept at it. I'm sure the government cut a lot of red tape to make that happen quickly as well, but the bottom line is, the lab strategies worked. If you can remember a time when the swabs were JUST a little larger than usual, those were the buccal swabs. Yes, I know they were a little uncomfortable, but hey, you got your test. Without the labs doing that, we all would have been sitting idle for quite some time.

The other labs weren't far behind, and they were manufacturing their own Covid swab kits overnight. Our practice went from getting a few test kits here and there to having more than we would ever need. We were now able to access them as fast as the lab could manufacture them. Those labs saved their own asses and, in addition, mine! Not to mention the countless sleepless nights they resolved for mothers, cousins, brothers, roommates, or whoever was "exposed" to someone for four seconds. If I could insert an eye roll, it would be right here.

In a worldwide pandemic causing massive supply chain issues, we could finally keep up with the demand for testing, which brings up another important point. While everyone was watching those Covid numbers climb every day, the figures were far from accurate. The massive climb in cases everyone was paying attention to didn't contain the actual number of cases out in the country; the figure represented the number of cases the medical community had the ability to catch up with. If practitioners hadn't shut down and additionally had access to testing supplies, the number of positive cases at the beginning of the pandemic would have been astronomical.

In other words, do you remember looking at that graph of the cases rising? That was just how many we could find. The real graph would have dwarfed the one you watched on the news. If you tested everybody at the same time early on during the pandemic, you probably would have just seen a newscaster throwing up their hands and saying, "Well, it seems we all have Covid. So goodnight, everybody."

CHAPTER 5

The Truth Starts to Leak

When the pandemic began, I had some major staffing challenges. My assistant, Alexa, who was with me for three years, was twenty-one years old at the time. She still lived at home with her parents, who were more than a little scared of what Covid might do to their family. Alexa called me one day to tell me that she couldn't come in because her parents would not let her out of the house and definitely not to go to work at a medical office and test Covid patients. I remember saying to her, "But THIS is what we do. If we don't do this, then who else will?" She seemed frustrated with her parent's decision, but I think she was also a little relieved because she didn't know what the outcome of the pandemic would be either.

I had the exact same conversation with my other assistant, Monique. She was a bit older than Alexa. However, she lived with her mother at the time. One day I got a call from Monique saying that her mother would not let Monique live in her house if she went back to work because she was going to catch Covid the minute she left the building. Even though we had some preliminary data showing that anyone and everyone had Covid in the early days, it was ironic that most of our patients who came back positive weren't even sick. Yes, most of our positives were asymptomatic.

I said to Monique, "You've probably already had it, and so has your mom! However, if you didn't catch it already and don't have it right now, you're going to get it at some point." Monique came back to work within a few weeks as she needed to support herself somehow; however, Alexa took most of the year off. This wreaked absolute havoc on my business as I couldn't operate as a one-man show, and trying to hire new staff during those first few weeks after the outbreak was a non-starter.

As a protocol, one of the things we kept in place—which, by the way, was completely unnecessary—was to call each positive patient personally to notify them of their lab results. We would record their symptomatology, and, in most cases, I had to talk them off a ledge. Most of my friends thought I was nuts for doing this. There I was, running around the State of New Jersey, trying to accommodate anyone who called in between driving, testing, coding, paperwork, and shipping. Sometimes I'd be on the phone with positive patients through the night and into the earlier morning hours. I would skip dinner or eat on the fly if I could find a restaurant or a drive-thru that was even open.

Like me, my staff was in the office every night until midnight, and I would order food for them, too, just so we could keep going. Some would work to get the requisition forms ready for the next day, while others populated the previous day's results into the patient portals. I had zero free time in my schedule, and neither did any of my staff. I felt bad for them, but what else could we do? While everyone else in the world was "staying home to save lives," we were helping to mitigate and understand this situation. Some nights, the lab results wouldn't populate until about 11 p.m., which seemed late to call and notify patients, but it wasn't fair to make our patients wait for information that we had in our hands, and most people wanted to know as soon as possible.

Our unwritten rule was that our patients would have their results the minute we had them in from the lab, no excuses. Whatever the time was, whenever the lab results populated in our system, I would call every single positive patient personally, one by one, that very night. Negative results were uploaded to the patient portals at the same time; however, in these cases, an automatic email report would be sent to the patient instead. It would have been physically impossible for me to call the negatives.

And just a quick word to all my patients who decided to call during the day asking whether I had their results yet: I am a man of my word! When I tell you I am going to do something, I DO IT. When I tell you I'm going to be somewhere, I AM THERE. Even though we notified every patient of this policy, my phone would unnecessarily ring every single day and night with the question, "I know you said they usually come in between six pm and midnight, but is there any chance you have them yet?" These calls really slowed the process and were unbelievably frustrating. Our days were like the movie *Groundhog Day*, doing the exact same thing over and over and answering the same exact questions daily.

At the beginning of the pandemic, everyone was scared. So, even though I usually made ten to fifteen phone calls a day to positive patients, I felt it was necessary to spend as much time as was needed on those calls, sometimes talking for thirty minutes or more. While most people thought I was nuts for calling every single patient, what my detractors didn't realize was that I was collecting very valuable data. This data was putting COVID-19 into perspective quickly, and I was finding a mismatch between what we were being told by those on TV and the real data.

A story I've told countless times that really helps put COVID-19 into perspective and gives you that thirty-thousand-foot view is from a home visit that I'll never forget. I got a phone call one day from a young man named Ryan. Ryan was twenty-one years old, and he presented with a cough, a fever, and extreme fatigue. He told me on the phone how sick he had been feeling, and I could hear his cough; he sounded horrible. Ryan asked me if I could drive to his home, about thirty minutes away, to administer a test. I rushed over to his house and used the exact same protocols I had in every house before his. I got out of my car, opened my trunk, and donned my mask and PPE. I called his phone and asked him if he would meet me out on his front porch. When he came outside, he also looked as horrible as he sounded, so I performed his test and got in and out of there as quickly as I could.

I went back to our office and sent his sample out along with everything else from that day's batch of testing. The following day, the results came in around 3 p.m., which was unusually early for the lab at that time. Ryan's name lit up like a Christmas tree in my report as all the positives showed up in a bright red color on my computer screen among the sea of black-and-white negative results. I called him to give him his positive result and went over our protocols. I also discussed his symptomology so we could record the information because we wanted to collect as much data as possible to make sure we could put the pandemic into perspective. Ryan was still febrile, coughing, and had horrible headaches. He still sounded sick on the phone but was optimistic that he would feel better in a few days. I reassured him that his exact situation was what I was seeing from our other sick patients, that after the fever broke, most people were recovering quite nicely, at home, without hospitalization, and in a few days.

Once Ryan and I ended our phone call, it was less than twenty minutes before my cell phone rang again. Ryan's mother was on the other end. She said, "I'm sorry to bother you, but you were here to test my son Ryan yesterday, and you just spoke to him to tell him he's positive. I hate to ask you this, but would you mind coming back so the rest of us can get tested? There are four more of us in the same house."

I asked her whether anyone else in the house had any symptoms, and she told me none of them did.

I replied, "Ma'am, I'm really sorry, but in order to get a test, you have to have some sort of symptom." She pleaded with me for testing as her son was still very sick, and the family wanted some peace of mind. They figured that if Ryan had it, then it was inevitable that the rest of them were going to get it as well. I agreed with her and figured this was completely necessary for science because the data would be invaluable. Since the lab had reported a little early that day, I was able to promptly head back over to the house in time to send out their samples with the day's batch.

When I arrived back at Ryan's house, I tested the family one by one out on the porch, just like I did the previous day with Ryan. As I was leaving, his mother came back outside and stood about fifteen feet away from my car, close enough to talk but far enough to ensure she was respectful of my space.

She said, "Ryan's been isolated this whole time, and none of us have gone near him. We should be OK, right?"

I reassured her that most of my patients were doing just fine and that we would all have answers tomorrow.

The following day, all four of the family's names lit up in red in the results alongside the word "POSITIVE." When I called to let Ryan's mother know the results, she gasped and asked whether I was sure. Then she paused and said, "But none of us have any symptoms. *Nothing*! I don't get it." She told me her husband had only been a little tired lately, and, other than Ryan, one of her sons had some diarrhea the day before, but it was gone now, and her other son had absolutely nothing. In this single household, the same virus and strain were present, but it came with completely different symptomology for each patient. In this five-person household, one was sick with "classic" symptoms, and the other four were asymptomatic or mild at best. That is, if you counted the diarrhea and slight fatigue as symptoms.

At that time, we all had no idea what Covid symptoms were, and we initially expected the symptomology to be consistent with that of the flu, like what her son Ryan had; however, this just wasn't the case. Now, if you are good at math, and I like to think I am, take that one story I just told you and multiply it by 441, because that is the number of house calls I made during this period. While everyone was locked down, posting "Stay home; it saves lives" or "I'm essential, I can't stay home," I was driving across the state, going from house to house, like a door-to-door salesman trying to make heads or tails of this.

The results from every house I went to were quite similar, and I was starting to doubt the doomsday prophecies the news outlets were predicting. In Ryan's household, we had an 80 percent asymptomatic grouping, with only 20 percent—Ryan himself—having any "Covid symptoms" at all. After adding up all the numbers from the other 440 homes I visited too, 10 to 15 percent of patients had classic symptomology, and 85 to 90 percent of patients were asymptomatic or mild. I can tell you that out of our over four thousand positive patients, only four were sent to the hospital. A fifth patient went on his own, but that was because his anxiety got the best of him. The hospital sent him home without treatment as he didn't require it, and it really doesn't affect the math much, but again, you need to understand our perspective.

Dr. Anthony Fauci would appear at White House Briefings to plead for the public to embrace longer lockdown periods. Fifteen days to stop the spread turned into twenty-one days, then twenty-eight days, and, well. . . . We all dealt with some form of lockdown, lockout, or disruption for longer than any of us would care to remember. The "experts" on this at the White House prepared the nation for doom and gloom. Quite frankly, they were scaring the shit out of people. The death count on CNN had to be one of the most irresponsible things I had ever seen from a media outlet. Now, more than three years on from March 2020, we've learned that the death counts were overinflated, miscounted, etc. I could have told you (and was telling my patients over three years ago) that most people were dying "with" Covid, not "from" Covid.

When I reflect on the totality of the circumstances, I think back to the assisted living facility that asked us to accommodate their weekly testing. I was thankfully able to acquire enough supplies and get enough staff together to take care of their needs, and this was where my perspective really began to shift. The facility had experienced roughly fifteen deaths over the previous weeks, and they were concerned about their remaining patients. The patients were all asking to get tested, as were the staff. When we showed up, the facility was set up how I imagined Fort Knox was. We rang the bell, and security identified us through a speaker system. We were only allowed in if we were wearing N-95 masks, and everyone in the facility was locked down.

My team and I sat down with the owner and his director of nursing to go over some protocols and then went into a large conference room to get our supplies set up. The mood in the facility was glum. These residents had been stuck in their rooms without activities or any social interaction. They were not allowed visitors, and after hearing of the multiple other residents who had passed, they were terrified. I had multiple requests for testing from the staff themselves, but the owner of the center made it clear to me that we were there specifically to test the residents.

As we walked through the halls with the help of some of their staff, one staff member started begging me to test her. She said, "It's nice that the owner is finally getting all the residents tested, but what about us? They forget that we all have families, and we are scared, too." She was very persistent and asked us to test her on multiple occasions.

I didn't really see the harm in helping her, so my team made a copy of her information to create a requisition form and took the sample. This all took place in the middle of May 2020. We had already been testing the public for months and this one test wouldn't detract from our duties at the facility. However, when I told the owner that we'd taken care of one of his staff, he wasn't happy. What I didn't realize at the time was how short-staffed he was; he didn't want to risk any possibility of losing another staff member for two weeks, which would put him in even more hardship.

When we received the results from the assisted living facility, fifteen out of the fifty-six residents were positive. That's 27 percent of the resident population. When I called the owner of the facility the next night to go over the results, his mood was somber. It sounded like all the air had been let out of him, and I could feel his emotions through the phone. These were seniors he had gotten to know by name and befriended. He knew their families, friends, and what their hobbies were. He had already lost fifteen of his clients, who he had considered friends, and there I was, giving him news that more of them may follow suit.

These were grave diagnoses for the residents. It was unclear whether they would develop symptoms in the coming days and succumb to this thing as well or whether the facility could sustain itself going forward. The owner wasn't quite sure what the state's policies would be on Covid cases in assisted living facilities, so he was also wondering whether he would be able to feed his family or if he would bring the disease home to them.

That same week, I got a call from an old friend, Mike. My buddy Mike had saved my ass when the place I had been living in was sold and, I'd had to move out. At the time, I'd recently started working for the Seaside Heights Police Department in New Jersey. Yes, that's the same town as the *Jersey Shore* show. No, it's not what it seemed like on TV. Mike, who was one of my fellow officers, heard I was having trouble finding a new place and, as a new officer, I wasn't making much money. He offered me a room in his house for rent at a very reasonable rate. Mike is a great guy, and after living with him for two years, I got to know his family well. They were very tight-knit, and it was nice to be surrounded by such a wonderful family.

Mike's sister Angela called me to tell me that their father was in the hospital with Covid and was on a vent. She asked whether I could take a ride over to their mother's house to conduct a test. "Of course," I said. "You are all like family to me; just let me know where and when."

Mike's mother was in her mid-seventies, and when I got there, she only wanted to talk about her husband. I tried to reassure her, letting her know about all the positive things I had seen in recent weeks regarding Covid, but Angela had told me that the prognosis did not look too good. Mike's dad was a tough guy, too! He was a retired United States Army Colonel. You wouldn't know it by looking at him, as he had to be about five foot five at best, but he was a bull and had a huge heart.

I did my best to console the family and took care of their mother as best I could. The next day, as expected, I called to give Mike's mother a positive result. She was completely asymptomatic and couldn't understand, after seeing what her husband was going through, why she was spared any illness. It made no sense to her that they were both infected, yet only one of them was sick. Unfortunately, I got the call a few days later that Mike's father had passed. I was heartbroken for his wife and entire family. I had known this family for many years and couldn't do anything to help save him. They felt helpless, I felt helpless, and the attention was now focused on Mike's mother.

Mike's family didn't want to get too close to his mother, considering her diagnosis, but they also had to conduct funeral services for his dad. Hearing of the way these services were conducted during this time would leave anyone scarred. Family members had to social distance and weren't able to see friends or extended family because the state prohibited gatherings over a certain number; it was pure chaos. Mike's mother never did develop symptoms, but she must now live with the fact that she couldn't be by her partner's side when he passed. She will have to live with that for the rest of her life.

CHAPTER 6

What Is A Virus, Anyway?

In my experience, most patients don't really understand pathogens and how they relate to their health. People simply visit the doctor or hospital when they are sick. We had never previously conducted any sort of mass testing on the entire population to see who had or didn't have a virus. We had only ever tested those who were sick to come up with a diagnosis and then treat them accordingly. Covid has since changed the way we test, treat, and respond to illness. There are some negatives and positives with this observation, which we will get to a little later in this book.

However, think back to early 2020, and all anyone wanted to know was the facts and figures about Covid, which people thought could potentially kill everyone on the planet. The CDC and those conducting the White House briefings should probably have just started with the basics. I'm convinced that if someone who knows all the *actual* facts were given just one hour on national television, everyone in the world would have understood COVID-19, and it would have prevented the years of uncertainty and bumbling response that we all subsequently witnessed.

So, here are the basics. Viruses are not living things. Well, the scientific community isn't 100 percent sold on this definition, but I can tell you that most scientists believe they are not living. We say this because viruses need a host to survive, kind of like parasites. Although a virus is not technically classified *as* a parasite, they are *like* parasites. In other words, they cannot get together and make little baby viruses on their own. They need a host to thrive, and, contrary to what one may think, a virus does not want to kill you because if you die, the virus dies with you. If a virus kills its host too quickly, it can't continue propagating, meaning it will burn itself out. There's a sweet spot in which a virus spreads rapidly enough to get a lot of people infected but not so rapidly that it burns through all its hosts too quickly to let it keep spreading itself.

Something people commonly ask about is mutations and whether they have a specific version of COVID-19. Honestly, I once had someone ask me whether they had COVID-19 or just Covid when we diagnosed them as positive. . . . Insert eye roll here again, please. Patients continuously came in asking if they had Delta or Omicron or something completely out of the ordinary. I appreciate the fact that the CDC is now using sequences of numbers and letters; this pandemic turned everyone into their own practitioner, and, in the process, it made everyone dumber.

In truth, it's completely irrelevant to ask whether you have a specific mutation. The proteins inside the virus's envelope do not mutate; it's the spike protein on the outside of the virus that makes that transformation. Imagine making a copy of a copy of a copy of something on paper using your photocopier at home. Eventually, the original image would become unrecognizable, but you would, however, see some weird shapes that develop on the copy. This is kind of what the spike protein does. It becomes so unrecognizable that the body doesn't remember it, and it can fool your system. However, once you get sick with COVID-19, your body develops a memory of most of the proteins inside the capsicum.

There are a small number of proteins inside a virus. COVID-19 has about twenty-seven of these proteins. While everyone seems to know—or at least think they know—what antibodies are, they don't often speak of the things that give you long-lasting immunity against Covid, or any other virus for that matter. This comes from your B and T cells. Your B cells are your "memory" cells, and your T cells are your "helper" cells. These cells are also often referred to as B and T lymphocytes. Once your body builds a memory of these proteins, your immune system may be able to recognize a given virus; however, the spike protein can mutate enough to fool it. Your memory of the entirety of the virus is what takes over once that spike protein gains access. You are now just getting a reaction to the spike protein. A good analogy would be that of a Trojan horse. The horse would be like the spike protein that gains access to the body but once it's inside and the proteins start to invade, you recognize them and can battle them much easier. This will be very important later when we discuss vaccines.

It is important to remember that viruses mutate from the moment they enter your system; they're looking for the best versions of themselves to survive. Again, they don't want to kill you, but viruses can and do make you feel horrible. The severity or lack thereof has to do with a few, and probably the most important, things I explain to patients. Again, you must see it from my perspective and a zoomed-out viewpoint; you can't use yourself as a data set and think you know everything about Covid. Lots of people do so and have their own personal perspective, but it's often completely wrong.

As of the time of writing, I got a call today from a patient who said she had Covid last week. She said that after she got better, she relapsed. The first thing I said to her was, "Are you sure it was a Covid relapse and not another virus?"

She replied, "Well, I had different symptoms from the original infection, but a friend of mine had the same thing, and she was positive for Covid."

To which I said, "How did she know she had Covid? Was it a rapid test or a PCR test? Did she go to a practitioner and get tested, or are you just sharing random information and trying to make important decisions about your own health that are based on nonsense?"

She paused and then said, "Yeah, I guess you're right." I talk to patients about things like this until I'm blue in the face, and once I think it's finally clicked, they call me back a few months later and ask the exact same questions. I think this has to do with what I call the "Fauci Effect."

I sometimes think back to a time when I had faith in Dr. Anthony Fauci and his team, who were assembled at the White House giving daily briefings during the pandemic. Here was this world-renowned doctor, director of the National Institute of Allergy and Infectious Disease, Chief Medical Advisor to the president of the United States, and a trained physician and immunologist. Who better to alert the country to the effects of COVID-19 and prepare the nation to deal with the fallout? Well, based on my data, he was incorrect in most of his assessments. Looking back now, my dog would have done a better job! The world would have responded better to this pandemic if someone stood there and said absolutely nothing.

Briefings were commonplace at the White House, and most of the nation was glued to them. Those of us in the medical community looked to them for guidance and any new data that would help put this virus into perspective. Accompanying Dr. Fauci was usually Dr. Deborah Birx. She was the Coronavirus Response Coordinator under President Donald Trump. Some of the things Dr. Fauci and Dr. Birx said would linger in our patients' subconsciouses for years to come, even after the data changed and proved the original hypotheses different.

This is what I mean by the Fauci Effect. It's almost every day that I find myself explaining COVID-19 to patients who were summarily brainwashed by the early reporting. It's as if everything they know about COVID-19 came solely from those early television briefings. Even though the information has evolved, most patients' understandings have not. It's the same as when a newspaper retracts a printed, front-page story due to sloppy reporting or new revelations. The public will only ever remember the headline and will never read the retraction buried on page twenty-six. This is why the Fauci Effect and the misinformation lingered long after new science proved the early science wrong. Let's face it; most people don't really pay attention to the news. They pay attention to headlines. People speak without listening to all of the data and get sidetracked by their own personal lives. I think the government officials speaking of the pandemic on television should have been far more cautious with their wording, considering the attention span of the populus. The entire situation was chaotic, and it was about to get worse.

CHAPTER 7

Masks. Yes, Apparently These Are Still A Thing.

In early 2020, someone asked Dr. Fauci whether the public should be wearing masks. I remember watching him during this exchange. Dr. Fauci blew the question off; he said that masks don't really protect people from viruses and that the public doesn't know how to wear them properly anyway. Too many people were touching their faces, which was the real problem because all that touching was partly to blame for how viruses transmit. In short, he said, *If you want to wear a mask, go ahead, but it isn't going to do much of any good for what we're facing.*

The media persisted, and so did the public. Therefore, the mask question was then posed to President Donald Trump. He stated that he would not be wearing a mask and moved on to other topics. As if a light switch had been flipped, anyone who was against the president started wearing a face mask, and the political virtue signaling began. You could tell what party someone voted for and what channel they watched just by seeing who was wearing a mask. Was the public doing anything beneficial in this instance? Let's look at the truth behind masks and put this topic to bed once and for all.

First off, we must look at the topic of masks from multiple angles. Face coverings are not all the same, and if you followed along on your own throughout the pandemic, I'm not telling you anything you hadn't already heard. The federal and state government instituted mask mandates around the country, essentially forcing people to wear face coverings in public. The FAA enforced that passengers should wear face coverings on airplanes. Some local municipalities even forced people to wear them outside while walking around. I got to see this first hand when I opened an office in Key West. A former police officer myself, I felt bad for the officers in Key West charged with enforcing this rule.

If you were walking down Duval Street in Key West and you weren't wearing a mask, an officer would walk right up to you and say, "Sir, you need to have your mask on. It's the law." *Really*, I would think. *I'm probably in the best place I could be,* outside, *and I still need to wear a face covering?* That was beside the fact that it was a hundred degrees outside, and I felt like I could pass out. I was in the police academy for six months and can promise you that there was no formal training on virology or epidemiology, at least at that time. I graduated in 2005, just to put that into perspective, but it wasn't that long ago! I thought it was insulting that government officials with no medical background whatsoever had come up with these silly rules like needing to wear masks outside.

The mayor of Key West enforced and came up with these ridiculous municipal codes, shut down businesses, and enacted curfews on the public. All that done, and she somehow still got re-elected! This is partially the Fauci Effect and partially the blatant stupidity of the residents and businesses who re-elected her. She's been proven wrong time and time again, but as you might recall from earlier this chapter, people only remember the headlines, not the facts. I promised no politics for this book, and I'm not picking a side here. I'm simply saying that this "leader" was proven wrong repeatedly, so the people who elect their "leaders" simply don't pay attention. I'd rather have someone with some common sense.

Here are the facts on masks: nobody who understands even basic virology would dare risk their life with an unknown, unstudied, deadly virus and use a simple face covering as their only protection. Virologists who work in viral labs wear sealed suits with air hoses for the single fact that one tiny, live viral particle is all it takes to infect someone. You can fit five hundred million COVID-19 particles on the head of a pin. Think about that. That's more than you can fathom counting. *Five hundred million on a pinhead!* However, this abundance of virus particles in your presence doesn't constitute a reason for not wearing a mask alone; it's about how viruses spread and how masks work. You see, particles behave differently the smaller they are in size. It's hard to put something so small and invisible into a tangible context, so we'll just work with the numbers here.

Any particle smaller than 0.25 microns is so tiny that it's not affected by gravity. In other words, those particles become part of air fluidity. For some context, you can't see air molecules, but you know they exist. Try fanning yourself with this book for a second. You can't see the air, but you can feel it around you. Particles this small float around in the air and bounce off things. They even bounce off each other. The problem with Covid particles is that they are even smaller than 0.25 microns. COVID-19 particles come in at a minuscule 0.1 microns, *two and a half times smaller* than the size needed to become part of air fluidity. So, consider someone already infected with Covid walking into a room. They breathe and cough out millions and billions of particles that float around the room. You are literally walking around in a sea of particles, and again, it only takes one single active particle to infect you!

When we discuss masks, we must be very specific. Remember, we are facing the deadliest virus ever known to humankind, right? Those surgical masks you saw everyone running around in are rated for three microns. Covid, again, is 0.1 micron, so it's thirty times smaller than what that mask is rated for. The problem isn't the size of the filter in those masks anyway, and, honestly, I hate even getting into discussions with people about masks because they always have some ridiculous rationale that isn't based on science or logic. The naysayers will say something like, "Well, I always wore a mask, and I never got Covid." But, as I always tell them, they may have gotten it and just never known. They could have been asymptomatic, just like the majority of my positive population.

Those surgical masks are never sealed to a person's face. Even if you duct taped the mask to your face, the filter still wouldn't be small enough to screen out Covid particles. Those little slats where the mask is supposed to pinch down onto your nose are also never completely tight, so air can get in here. That's why people's glasses fog up when they are wearing a mask. Air will always take the path of least resistance, and you essentially create a venturi effect around your nose where the air rushes in and out quicker. Ever seen a drywaller wear a respirator when they work? As soon as they take the respirator off, they still have a ring of white around their nose, and you can SEE those particles.

Then, the argument becomes, *Well, it traps some of the particles, which must help, right?* Nope, study after study has shown that most of what gets trapped in masks are larger globs of dead particles. The live virus is the tiniest of particles. It's like going into a coal mine with a surgical mask on. When you walk out, you're going to have coal dust all over you, in your nose, inside your mask, and, more importantly, in your lungs. Remember, it only takes one Covid particle for you to get infected. I feel like I need to reiterate that to my patients because since they can't see it, it doesn't exist. Most people like things that are tangible to them, or they feel like they are just relying on magic. What's worse, we have more and more patients who have developed other respiratory problems from wearing masks because they trap moisture and moisture breeds bacteria. You are making yourself sicker by wearing that thing all day!

So, what about N-95 masks, you ask? N-95 masks are typically rated for 1.0 micron. Remember, COVID-19 is 0.1 microns, so the filter on the N-95 is ten times too big to filter out the virus entirely. N-95 masks are your best bet, though. They are an easily accessible, commercial option to protect yourself from COVID-19 or any other respiratory pathogen. However, they are never worn correctly, and even when they are, it becomes useless in the moments when you scratch your nose and break the seal the mask has on your face. Do you really think a loosely fitted N-95 mask would be a fair fight for the single particle that is required to bind in your system?

Remember, viral particles transmit as part of air fluidity in a closed environment. Study after study has shown that no number of wiping things down, sanitizing plane seats, etc., will do any good at protecting you from acquiring this virus if you are exposed to it. Don't get me wrong, simple hand washing and not touching your face is a great way to inhibit the spread of disease, but in a closed environment, you'd better be wearing a sealed plastic suit. If not, you are as vulnerable as anyone else.

Most viruses are similar in size, though some are much larger, and bacteria are huge when you scale them up against a virus. Just think about all those bacterial infections you're inviting into your system when you have that hot, moist cloth around your face all day. Even though properly worn N-95 masks are your best bet against infection, the governments around the world who enforced the mask mandate failed to mention this to the public. If you believe you require a "face covering of any kind," you have lost credibility with those of us who take facts, logic, and data seriously. When I went somewhere that required precautions, I wore a properly-fitted N-95 mask; I never touched it once it was on my face, and I took it off properly, although I knew my chance of getting infected was very high.

If, when Dr. Fauci mandated masks on planes, he had said only N-95 masks would be permitted, I would have had a lot more respect for the government's effort to mitigate this virus—but they didn't. In fact, they just continued to insult all our intelligence by waffling on about masks and requiring you to wear anything that covered your nose and mouth. So, basically, you could have worn pantyhose over your head and gotten onto a plane. If you had tried this before the pandemic, I'd be sure you were trying to hijack my flight! I'm not going to bother mentioning bandannas or cloth coverings because, at this point, if you don't get it, well, you're probably waiting for Dr. Fauci's biography to come out. Many positive patients told me that they "just don't understand how they could be positive" because they have "been wearing a mask the entire time." Insert eye roll again because that's about all I could do on the phone besides letting those patients know that masks just aren't effective.

My mother had called me early on during the pandemic. She was scared to leave her house and asked if I would run out and pick up some groceries. She only lived about thirty minutes from me, so I wrote down her shopping list and headed off to the store. When I showed up at her house, I rang her doorbell. I stood there ringing and knocking for a few minutes until I heard the garage door open. I walked off the porch and was met with yelling, "Stay back," she said. "Put the groceries by the back of the car. I need to wipe them off!"

I obliged, but once I got a glimpse of her, I started laughing. She paused and said, "What's so funny?"

I replied, "Well, first of all, if you are that worried about this virus, maybe you should start wearing your mask the right way." She was wearing a surgical mask; however, it wasn't even on correctly. It was upside down; the metal pinch bar was under her chin, and her nose was fully exposed as she kept grabbing it and trying to pull it back up. I left so she could get her groceries disinfected and get back to the CNN death count that had been prevalent on her TV screen for months. Thanks, CNN, for scaring the crap out of my mother and turning her into a hermit; she still barely leaves the house three years later.

We never *ever* required a patient in our offices to wear a mask upon entering. If they wanted to, that was their prerogative, but the thing that always irked me was when patients felt it necessary to try and school me on mask usage. Here I was operating a medical clinic with trained medical staff when a patient would walk in and start rudely asking: *Why aren't any of you wearing masks?* Look, lady, what do you do for work? Oh, you're an accountant, that's nice! Do I come to your place of business and tell you how to fill out my tax returns? When was the last time you took a class in virology?

It's insane how this pandemic turned everyone into their own practitioners. As if self-diagnosing using Web MD wasn't bad enough. Practitioners hate it when a patient looks up their symptoms and assumes their diagnosis before even consulting a professional. I have always told my patients, "We do smart medicine. We're not going to tell you to do something because it makes you feel psychologically better; we are here to actually help you feel better."

Using inferior masks for a virus that simply isn't affected by them is the second stupidest thing I have ever seen in my lifetime. I'll cover the stupidest thing later in chapter twelve. Scientists know the relevance of masks when it comes to Covid, yet they ignore facts to further their own narrative. This, again, was a result of the Fauci Effect, even though Dr. Fauci had it correct to start with. The Surgeon General of the United States, Dr. Jerome Adams, said the exact same thing as Dr. Fauci at the beginning of the pandemic, that masks make people touch their face even more than usual, the public doesn't know how to wear them correctly, and they give people a false sense of security. The only thing that really works to mitigate viral transmission is distance and ventilation. When people wear masks, it also makes it harder to understand when they talk. Then because no one can understand a thing they're saying, people get even closer and . . . well, you know. Clearly, social distancing at its best! *A government policy that had the opposite effect of what it was intended to do – it can't be!*

It's sad that health professionals went along with this narrative for so long instead of just citing the science and telling their patients the truth. The fact that this happened has cost our healthcare system its credibility. If anyone actually paid any attention, the CDC would have the most to lose. The three studies listed on their own website citing mask effectiveness had little to no participants, and most were self-reported from early on during the pandemic. They used shoddy science to back a narrative that could be easily dismissed, yet rather than dismissing it, the Fauci Effect took over. If a grown adult truly believes they are invincible against viruses because they're wearing a mask, I blame the adult for their lack of research because the science is out there. However, most adults don't know how to conduct research or where to look if they want to. This is why I must place most of the onus on the CDC; they are very culpable in this instance.

I like to research, and I want my points to be salient when I disseminate information. It is very important that my staff and I remain credible because once you lose your credibility, everything else goes with it. Early during the pandemic, I liked to cite an associate professor at MIT. Her name is Dr. Lydia Bourouiba. She conducts studies of fluid dynamics and has written many papers on the fluid dynamics of disease transmission. In one of Dr. Bourouiba's studies, she used high-speed cameras to observe people breathing in a room with no ventilation. Her conclusion was that particles traveled up to 27.2 feet.

While six feet is a nice, easy number to remember and sounds keen, the idea that you're safe from Covid if you're six feet away from someone is made up. It's another fabricated, scientifically insignificant "fact" that Dr. Fauci and his team would have you believe. I want to reiterate that it's more important to tell people the truth and give them the proper tools rather than making things up that sound good just to gain everyone's compliance. It's ridiculous and absurd that these people at the top couldn't treat us, the population, with dignity and respect. Instead, they chose to run with a distinct narrative that most people still believe today.

When the mask mandates and the federal mandate for domestic air travel were finally lifted, most people stopped wearing masks. However, they remained an avenue for virtue signaling. I traveled all over the country and will tell you that the percentage of people in Florida wearing masks was much lower than in an area like Los Angeles. I flew out to LA for a wedding in May 2022. My buddy Kato Kaelin was getting married; yes, *that* Kato Kaelin from the OJ Simpson trial. I was lucky enough to get into first class on the way home, and out of the roughly forty-four pods in first class, I think I was one of two people not wearing a face mask.

I can't tell you how many times I've pulled out a surgical mask and my vape pen in my office to prove a point. I would take a drag, pinch the mask as tight as I could to my face, and blow it out. When my patients saw the vapor flow through and around the nose area of my mask, they usually took their masks off and threw them in the garbage! Sometimes, you must make the facts tangible for people.

Again, people will say to me that wearing a mask has got to be better than nothing, that it has to give you SOME protection. To which I say, "Unless you are wearing a properly-fitted N-95 mask, you will get absolutely zero protection." This was besides the fact that so many other pathogens could be growing inside that thing, which could make wearing a mask much worse than going without. People generally reply by asking why doctors in hospitals wear them. However, doctors and nurses wear masks in hospitals to not get splatter on their faces, to not drool in the operative field, etc. If you are comparing apples to apples, one should ask: *Why don't virologists wear surgical masks in virology labs?* The answer is BECAUSE THEY DON'T WORK FOR VIRUSES. Period!

CHAPTER 8

We're Doing What Now?

I need to add some dialogue here about some of the other PPE that was being used during the pandemic because the whole thing about wearing gloves made me rethink the average IQ of the nation as well. One of my buddies commonly says to me, "The average IQ in this country is one hundred, so think about that. HALF the country has an IQ of LESS than one hundred!" To demonstrate, in April 2020, I was on a flight from New Jersey to Florida. I was in an aisle seat, and there was a woman sitting one row ahead of me in the opposite aisle. She was wearing what looked like a hazmat suit and had a mask on. She was also wearing blue latex gloves, which were a common sight throughout the pandemic. However, she was holding and reading the newspaper with her gloves, all while digging deep into a bag of popcorn throughout the flight.

I don't think regular people realize that when practitioners wear gloves at work, the gloves are contaminated as soon as they touch something. That's why practitioners change them between every patient. We don't read the funnies with them on and eat our lunch. I have a buddy who won a Superbowl with the Colts in 2007. He came over to my house during the pandemic, and he was wearing a pair of gloves when he walked in. As soon as I saw this, my eyes rolled into the back of my head, and I said, "Dude, why are you wearing those?"

He said, "I'm just trying to stay safe!"

My reply was, "Really, how long have you been wearing those today? When was the last time you washed your hands—I mean, your gloves? Take those dirty things off, and please just go wash your hands!"

He looked at me, laughed, and said, "Yeah, I guess you're right. I haven't washed them all day."

It's kind of comical if you think about it. In every sort of specialty, professionals of all types, are tasked with understanding the jargon, using some sort of unique gear, or just learning the ways of the job. Here we had an entire population who now thought they were "educated" on virology, epidemiology, immunology, etc. Arguing with them was useless because most had never listened to the facts about Covid; their arguments were usually just rhetoric. The face shields really made me laugh. Seeing these idiots walk around in an airport, airplane, outside, inside, or wherever was just hilarious to me. In fact, the woman in the hazmat suit with those gloves also had one on. The only thing she was protecting herself from was humiliation because nobody could see what she looked like under all that "protection". Maybe she was shy?

The face shields also reminded me of those plastic partitions some offices set up. I walked into my attorney's office in Key West one day and wondered if I had hired the right guy. His front office is a big open area with no half walls or partitions. His front office receptionist sits behind a desk in this big room. He'd set up a two-by-two partition, so when you walked into the office you were in a completely open area, but as soon as you sat down, this tiny partition was in front of you and her, kind of like those face shields. If I moved my head six inches I was now breathing on her side of the plastic.

Approaching him, I said, "I'm starting to worry about this office. I thought I was coming here because you all were intelligent and I was going to the best and brightest! Why on earth would you have THAT over at the front desk?"

He told me, "Yeah, I know, we just have it there to make people feel better."

My reply was, "Really? So you know it doesn't work, but you keep perpetuating this nonsense because you want to make people FEEL better! Well, you are insulting my intelligence in the process!"

This seemed to be the running theme during the pandemic; placating others to avoid being called out on social media or some other platform for not abiding by the "rules." To those people who did that, I say: grow up and worry about yourselves. I don't understand where every single Tom, Dick, and "Karen" who tried to enforce the government's guidelines came from during the last three years. Some people clamored for a lockdown, to keep kids out of school, to tattle on anyone without a mask or who wasn't socially distancing. Well, none of that worked anyway, so congrats on just making people angry. None of you were helping keep anyone safe!

Once I had some data in early 2020, I said that we should adopt the Swedish model for the pandemic. I put a post up on Facebook and was instantly ridiculed by anyone and everyone saying, "What is wrong with you? People will die! Is that what you want?" Keep in mind, these were people from all different walks of life, but not one of them was an expert on Covid or had any actual data on it. These were all people who had an "opinion," but their opinion didn't change once new data came to light. Most of them were the same people who still wear masks on planes today!

If you look at the curve of cases in the United States, it goes up and down. There is a reason for this, and that is because there is a seasonality to the virus, just like the flu. However, the seasonality of COVID-19 is different in the Northeast as it is in the Southeast and so on. After our first year of gathering data, we were able to predict when cases would rise and fall and in which region. The only thing we were uncertain of was how much. This had to do with access to testing and how it was being reported. Once you understand testing, you'll see how the different kinds of tests measure up against one another. However, *all* PCR testing gets reported to government agencies by statute. This means that the positive count you saw was relatively accurate, considering the shortage of testing supplies. When one of our labs tests an individual, they must report it to a state agency, which then reports to the county level. When a rapid test is conducted at home, those numbers don't get reported anywhere! This is another story. . . .

We were becoming experts at transmissibility in certain areas across the country throughout the process, and I would use this data to alert my staff in advance as to when to stock up on supplies. In August 2021, my Florida staff was becoming overloaded. Therefore, I made sure I was stationed at that office that month to ensure they had an extra pair of hands. One of my staffers begged me to hire an additional person, which in normal circumstances would have been warranted. I said to her, "We just need to weather this storm for a few more weeks. I don't want to train someone just to lay them off at the end of the month." My staff thought I was just being cheap, but as soon as the month ended, our COVID-19 cases and patient visits took a nosedive. We went from seeing over a hundred patients per day to under twenty-five per day overnight. The same was true for the seasonality of Covid cases in New Jersey. Once we got through December and January with similar numbers every month, February was like a post-apocalypse. Where had all the patients gone?

Common sense disappeared during the pandemic. Usually, what brings people together makes them stronger and smarter, but in this instance, it drove them apart and made them dumber! I was on a flight from Houston to Newark one night after working all week in my Key West Clinic. I was upgraded to first class, which was nice, and I considered taking a nap because I was exhausted. When the flight attendant working for the first-class cabin came on the PA to give his "COVID-19 briefing," it really irked me. Day in and day out, I explained COVID-19 to patients and explained the hard science to them. I did this every day. So having a condescending airline employee talk to me like I was a child and explain what I had to do to "stay safe" was the last straw. Not to mention that he was wearing GLOVES! I had to say something just to let him know he wasn't the smartest on this flight when it came to Covid.

I told him, "Why in the world are you serving me my food with those gloves on? They are contaminated the moment you touch something." He asked me how I knew that, so I introduced myself and told him I ran multiple clinics and about the thousands of patients and thousands of tests we had conducted. When I thought he was going to come back with something snappy, he asked me whether he could pick my brain instead.

"Sure," I said. "As soon as I finish my meal and you are free, come and grab me."

He came back thirty minutes later and took me up to the front of the plane. In a quieter setting, away from the passengers, he said, "Can I show you something?" I nodded, and he proceeded to take off the latex gloves he had been wearing. His hands were RAW! They had red blotches all over them and were oozing fluid. He asked me what he could do to fix this.

I said, "You can start by taking those ridiculous gloves off." This guy would wash his hands and then wipe them with alcohol sanitizers before donning gloves. He would wear these gloves every shift and probably in between flights. That and the hand sanitizer really did a number on his hands.

We spoke for about forty-five minutes. I shot straight with him on gloves, masks, transmission, and so on. I gave him the same conversation I had with my patients daily. I even convinced him to take that stupid mask and his gloves off. It felt good to be able to help this man get over his fear. His name was Mike, and I'll protect his identity here, but Mike, if you are reading this, I hope your hands cleared up. Mike's problem wasn't just his hands; it was that he was a prisoner in the narrative. I truly hope our conversation helped him to enjoy life once again and get over the fear that certain media outlets put him in. Shame on them for doing that. They should have consulted people with real data rather than focus on scaring everyone to death in the name of ratings.

On the other end of the spectrum, there was the opposite of people like Mike, who didn't care about anything in the world, least of all Covid. I was in my Florida office one day and overheard a conversation between a patient and my staff. The patient was about twenty-one years old and almost spoke in that California Valley voice, or what you'd hear in an '80s movie about surfers. He kind of sounded like Keanu Reeves in Bill and Teds Excellent Adventure. He was coughing, sneezing, and sounded awful at the window while talking to my office manager.

He said, "I've got to get tested for Covid. The fast one—the one that comes back right away."

My office manager said to him, "Sir, you don't sound too good. Our office can send out a full panel to see if there is anything else lingering so we can treat it properly."

He then replied, "*No*, I just need to get tested for Covid so I can go back to work."

By the way, people like this patient are your neighbors, your friends, your coworkers. They don't care if they give you the flu or Respiratory Syncytial Virus (RSV); they just want a Covid negative so they can return to work. Real smart, right? I told my staff I would take this one and would be happy to explain things to him in the exam room. When I walked in, not having seen him earlier at the window, he was wearing a very interesting shirt. His shirt said, right across the front, in big letters, "Professional Raw Dogger." For those of you who don't know what that means, I encourage you to look it up. My thought to myself at that moment was, *He should probably be getting testing for something else!*

That patient came back negative on the rapid, which meant absolutely nothing. We couldn't find out what he had since he refused to get tested, so we provided him with a note that said he tested negative for COVID-19. I'm quite sure he went back to work, and I'd also bet that he infected at least half of his coworkers with whatever he had. Common sense went out the window on both sides of the aisle. Young, old, in the middle . . . it didn't matter. Patients demanded certain things because they thought they knew the system better than you. They would walk in demanding full panels without Covid symptoms; you can't do that! They would cherry-pick Covid-only tests just to get their desired result. This was less about health care and more about gaming a system. Considering there were people susceptible to getting ill from this thing, I think many people acted irresponsibly with their carelessness. I also believe a multitude of people completely overreacted by using selective outrage for a by-line that fit their own narrative.

Today, at the time of writing this, in April of 2023, I had a patient miss an appointment. She tried to blame it on her daughter having a doctor's appointment at the same time as her appointment but in a different location. Well, we see kids in our office, so she could have killed two birds with one stone. She was probably one of the people who would get a negative rapid to overturn a positive PCR to be able to take a trip to Aruba, except here she was trying to get out of paying a fifty-dollar cancellation fee because her daughter had "Covid!"

She said to me, "Well, I'm sure you wouldn't want me bringing my daughter to your office and exposing everyone to Covid."

I told her, "You probably should have since we are the experts in it!"

My point is: Covid has become a convenient excuse for people who don't want to take responsibility for their mistakes. The hypocrisy drives me nuts, and it's the government's fault for starting this rollercoaster by disseminating erroneous information. They kept the ride going by not correcting the narrative once the information was considered null or verified. They fed news outlets and people's fear and destroyed any semblance of common sense that was left in the world. People could now use this excuse or the fear of losing wages, not being able to travel, or whatever drove them for their own benefit. They would lie, cheat, and steal so they could game the system or use it to their advantage. It was tasteless and still is.

CHAPTER 9

How Testing Actually Works. There's More To It Than You Think!

When the pandemic first started, our lab was already using PCR testing for our RPPs. "PCR" stands for Polymerase Chain Reaction. This type of test looks for a signal and amplifies it thousands, millions, and billions of times to isolate it and determine if a patient is positive for a specific pathogen. As of writing, our lab tests thirty-one things on its panel, which vary between viral, bacterial, and fungal pathogens. These pathogens are all very different and don't all respond the same to antibiotics. For example, one fungal pathogen we test for requires its own unique course of treatment. These pathogens can also cause different symptomology, but humans tend to respond typically to most of them. Everybody knows when they have something brewing in their system. A runny nose, cough, fatigue, aches, and pains are all signs that you may be "coming down" with something.

The world was clamoring for a rapid test. Abbot was the first to come out with a PCR rapid test, but it was expensive and not easily accessible to the public. These were typically found in hospitals and could provide a result in about thirteen minutes. However, outside of taking a trip to your local hospital, there was nothing available in a household environment that could give people ease of mind and a quick diagnosis. Eventually, companies started to manufacture the rapid antigen test, which for some was a blessing, but for those of us who understand the importance of testing and getting a proper diagnosis, it was a complete nightmare.

It's important to know how testing works to understand how to diagnose COVID-19. PCR testing is the ONLY proper way to diagnose COVID-19, as other methods can be wildly inaccurate. When the pandemic started, we only had PCR testing available; a rapid antigen test had not yet been validated. Most people had to wait seven to ten days for labs to process their results, and the waiting was painful. Our company was in a unique position. We got our results back from our boutique labs within twenty-four to forty-eight hours from day one. We had become accustomed to the twenty-four-hour cycle as the labs would rush RPP results to us as quickly as possible. We would collect the day's samples and then ship them overnight via courier, and the lab would receive them the following morning. Once the lab had them in hand, they would take a couple of hours to run, and we would have the results in our portals as soon as they were reviewed by the PHd at our partner lab. We would then promptly notify patients of any results and make medical recommendations. Why the larger labs were taking seven to ten days was beyond me. Eventually, those larger labs started getting patients their results in about three days, but that was completely unacceptable to me.

Once the pandemic started, business as usual was disrupted because of the added pressure on our labs. This pressure caused severe backups and slowdowns on the status quo, and we all needed to adjust accordingly. Some of the biggest issues we had were with the shipping companies. See, once we take a sample, do our job, and ship it off to the lab, the latter half of the process is completely out of our hands. Therefore, if there was a storm in the middle of the country, UPS and FedEx would experience major disruption causing massive delays. We once had a shipment of over a hundred samples delayed for over a week due to storms. Patients would start calling and freaking out on my staff, but there was really nothing we could do. We had to put notices up in our patient privacy practices and make people sign a waiver that they were aware there could be delays in receiving results due to uncontrollable factors. When people are frantic, they will blame whoever is closest to them, which wasn't fair to my staff.

Some patients were getting tested for Covid out of necessity because airlines or certain countries wanted testing conducted within a certain timeframe before boarding. Most of the world started to travel after the initial scare, and I'm sure most of you had to upload a negative PCR test into a flight portal with a seventy-two-hour window. Our company was the most reliable entity to handle PCR testing in both our New Jersey and Florida offices, and if the shipping company screwed up, we heard the brunt of it. While 99 percent of the over forty thousand tests we conducted during this period came back within twenty-four hours, the tests that did not were a major cause of stress for my staff and myself. We would recite the EXACT same verbiage to patients REPEATEDLY. Our Florida office dealt mostly with travelers coming in and out of Key West—these were often international travelers. While we had a stellar reputation for being the fastest testers on the island, when shit hit the fan and test results got delayed, people were quick to blame us and us alone.

Every patient was told, "We never guarantee anything because once we ship this, the sample is out of our hands. A typical turnaround is twenty-four to forty-eight hours from the date of testing, but we DO NOT assume liability if the shipping company or the lab has any delays. We are NOT the lab, nor are we FedEx or UPS." I would then add, "If a UPS driver decides to get drunk and roll their truck into a ditch or the FedEx plane goes down like in *Castaway* with Tom Hanks, DO NOT CALL HERE asking how we can speed up your results. No amount of calling, writing, texting, or smoke signals will make that happen. We put your results into your portal the MOMENT we have them back!" This was the absolute truth, and we walked through fire for patients. My staff did a tremendous job during this trying time, but angry patients sometimes get the best of you.

Having my staff repeat verbiage constantly came in handy when I got off-site calls, and the patient said something along the lines of, "*Your manager guaranteed that I would have my sample by this afternoon!*" Now, I don't love confrontation, but don't lie to me! I know EXACTLY what my staff told you and EXACTLY the way they said it. Even though my staff and I work as a team, sometimes certain patients think they can get one over you. They are litigious and look for reasons sometimes to jam you up. I did feel for these people; they were caught between a rock and a hard place and just trying to navigate the many restrictions. However, I had a lot of patients nicknamed "Karen" during this period. Nevertheless, 99 percent of my patients were great. My staff and I made a lot of friends along the way, many I now consider my best friends in life, such as Carnella and Joe.

I had just opened my Key West office in November 2020 and got a call from a woman named Carnella within a week of opening. Carnella's husband, Joe, then came in for a Covid test. I wasn't very busy at the time, and Joe and I got to talking. We were on the same page, and Joe was very interested in the data we had collected on COVID-19. After a half hour, we exchanged phone numbers and made plans to meet up. People like Joe made all the challenging patients worthwhile. Joe and Carnella are two of my most trusted friends today, and Carnella is a very accomplished author who indirectly inspired me to write this book!

Having to wait and rely on outside factors to make sure you could catch that flight home was never fun, but my staff and I did our very best. My staff got so good at tracking samples that I would find out about any delays first thing in the morning. I would know where the samples were stuck, and how many and relayed that information to the lab so they could start hounding the shipping companies. The labs were paying for overnight, first thing AM delivery, so one would hope that the large shipping companies would put extra effort into making sure those samples were delivered in time. If a sample was shipped from Key West to San Antonia but got stuck in Ft Lauderdale, you would think it got priority the next morning because it was originally sent overnight, right? Nope, somehow, they got lumped in with the ground packages, and the sample could sit there for days. The shipping companies were a huge headache during the pandemic. It reminded me of the saying, "We're all in this together." Clearly, the shipping companies hadn't caught on! There is a private UPS store in Key West that wouldn't even accept our packages; the manager would refuse them for fear of Covid because the boxed-up samples said "laboratory samples" on the outside. This was totally inappropriate of the store, but it gives you an idea of some of the challenges my company faced during the pandemic.

I remember one patient who gave my office manager such a hard time in our Key West office that my staff member called me crying. While I like to think I'm a nice guy, I am also very protective of my staff. I had seen them work so tirelessly over the course of this pandemic. They would stay up all hours of the night to pitch in for calls, lab results, and whatever else it took to make our patients happy. I was NOT having it when someone didn't understand our perspective and wanted to try and strong-arm one of my employees. I found out who it was, called him, and told him how inappropriate it was for him to take out his misplaced anger on this young woman. I won't repeat what I said to him on the call, but after we hung up, he called her to apologize. Like my good friend David Richter always says, "Business would be great if it wasn't for customers or employees." Too bad you need them both to stay in business!

Once available, we initially refused to use rapid antigen tests in our offices because they do NOT provide a proper diagnostic for COVID-19 and sometimes take you down a road you think looks similar but lands you on another planet, for lack of a better reference. Once you look at the science of these antigen tests, you'll understand why, as a reputable medical office, we wanted nothing to do with them. Nevertheless, as more and more people asked for them, we caved and decided to offer them out of our Florida office. I've got to run a business, too! The office down the street was conducting ONLY rapid testing and getting over a hundred people coming through their doors per day. While I thought this was ridiculous, I eventually decided that I needed the work more than my pride.

Let me put testing into perspective for you. You'll have to use your imagination a bit here to picture what is going on. The sensitivity to PCR testing is logarithmic, not linear. In other words, the sensitivity levels don't go in order like counting from one to five does. The sensitivity power is increased by the tenth power at every level. PCR testing starts at a sensitivity of one, then goes to ten, a hundred, a thousand, ten thousand, a hundred thousand, a million, ten million, and so on, and so on, all the way to level thirty-six. It's mind-boggling. Basically, if you're looking for COVID-19 at level one, you are looking for particles in an area the size of your cell phone. At level thirty-six, it's like searching for particles in the entire solar system. At that highest level of thirty-six, you only need nine copies of a virus present at the lab level to make a determinative positive call and say that it has bound and is in your system.

In comparison, an antigen test is very different. You would need 106 to 109 of a virus present to turn that little rapid card blue. What does this mean? Essentially, instead of nine copies of a virus present, one would need nine million to nine BILLION copies for the test to come out positive. So, if your rapid test turns positive, you are POSITIVE for "something." Those rapid tests can turn positive for any coronavirus. There are seven that we test for on our RPP, including Human Coronavirus (the common cold). To sum up, if a rapid test comes up positive, it tells us ZERO scientifically, and it tells us even less if it comes up negative.

Remember that testing center we were competing with in Key West that was only conducting rapid tests? I'm quite sure they had a lot of positive people running around the island who were convinced they were negative. We saw this all the time once people started caring less about the pandemic. Someone would come in for a test on a Monday. I would call them Tuesday night to confirm the positive result and go over protocols. Then, later, I would see them out around town, and they would tell me something like, "Yeah, well, I went and got a rapid test right after you called me, and that came out negative, so I went back to work." Why bother getting tested if you were going to ignore the results in the first place?

Covid really changed the world in many ways, and not for the better. I can't even recall how many arguments my staff had with patients. There were many facilities that hired us to conduct weekly testing, so our perspective was a little different from most. When you test the same people over and over, every single week, you collect a lot of relevant data. There was a municipality that we derived an enormous amount of data from between 2020 and 2023. In fact, at the time of writing, they are our last remaining weekly testing partner. While the number of individuals has dropped significantly since the pandemic has calmed down, we used to see over one hundred individuals each week at that facility. I had to call, at least ten police officers each week to notify them of their positive results, but I had to convince nine of them they had it! In other words, most of them were completely asymptomatic. This is the point, as someone testing people in mass, where you really start to notice things.

You see, COVID-19 isn't just about a simple negative or positive result. You are negative, or you are positive with caveats. There are three major caveats, to include, viral load (measured in CT value), co-infection and co-morbidity. CT value or Cycle Threshold indicates the level of sensitivity needed to detect the virus using PCR testing, with each level representing a tenfold increase in sensitivity. As the CT value increases, the number of virus copies needed for a positive result decreases, with just nine copies of the virus making a positive call at the highest level (level thirty-six).

Here's what we learned: most people who come back between level twenty-seven and level thirty-six didn't have a single symptom. If someone came back at level eighteen or nineteen, they would almost definitely have symptoms. This is exactly what level my own sample came back with when I got Covid in October 2021. However, higher levels with no symptoms were the norm—remember that 85 to 90 percent of our patients were completely asymptomatic. Most people expected that if they came back positive, they would have some kind of symptoms. This simply wasn't the case. Having to explain this to patients repeatedly was a real pain. They would think that the testing was bogus before listening to the data because the news was painting a grim picture of what happened to people who contracted the virus.

I'll give you an example of one of my more challenging patients. One of the police officers who was subjected to weekly testing came back positive during one of the weekly testing cycles. This officer and his wife and daughter had been to my home multiple times, and I treated him no differently than I would my friends. So, I was taken aback by how defensive he was on the phone when I called him to give him his results. He said, "There's no fucking way, Mike. I don't have ONE symptom; it has to be a false positive," and he claimed he was going to lose out on a full weekend of overtime that he desperately needed. I told him that false positives were almost impossible. You can only get a false positive if there was spill-over at the lab or the sample was contaminated, and we took great measures to make sure that this didn't happen. The officer then asked if he could come by my place to get a confirmatory test the next day, to which I obliged.

When he came by the office the next day, I tried to explain the science behind testing and that he was actually "in the norm." Nevertheless, we tested him and waited patiently for the lab results. The following evening, I couldn't wait to open my computer screen when I received the notification that the results had populated. There it was in all capital letters: "POSITIVE."

The lab assigns each case an accession number when it arrives without differentiating between patients or dates of birth. Samples are either labeled POSITIVE, NEGATIVE, INCONCLUSIVE, or QNS. While I'm sure I don't need to explain the first two categories, I'll explain the latter two for clarity. An INCONCLUSIVE sample means that there is a signal, but it's not strong enough to warrant a POSITIVE call. This result typically indicates that the person is in an incubation period or at the tail end of the virus. A second test would determine whether they were waxing or waning. Many INCONCLUSIVE samples would turn up POSITIVE if retested a day or two later. QNS simply means Quantity Not Sufficient, which can occur if the sample leaks during transport or is not large enough for testing. While this rarely happens with RPP or Covid samples because of their size and packaging, it's a common issue with urine samples for other types of tests.

When I called the officer again to inform him of his POSITIVE result, he was slightly calmer than before, but he still gave me a ration of crap about not having any symptoms. I asked him, "What do you think the symptoms of Covid are?"

He quickly replied, "Cough, fever, shortness of breath." I told him that while those are symptoms, they're not the most common ones we see. I then asked what he thought the most common symptom was.

He didn't have an answer and paused before I interjected, "IT'S A HEADACHE!" Early on during the pandemic, there was a study out of Germany that had to do with hemoglobin uptake issues, which is why most people who get Covid develop a headache. This is the most common thing we see, and when you study the metadata, the picture becomes clearer.

Just a few days later, I was going over our results to start making routine nightly calls to positive patients, and that officer's name was on the list again! I called the officer and said, "I'm hoping a third test a week after your first test finally convinces you that you are positive. Can you please just get through the isolation period and not subject anyone else to getting sick?" He hung up on me but first threw in another jab about NOT having any symptoms.

This scenario was all too common after rapid tests became widely available. People would freak out when they came back positive on a PCR and run to the pharmacy to grab a home test just to negate our findings. Now that you are well-educated on the different tests, you understand that this officer was indeed positive. The second caveat with a positive case is co-infection.

Most sick patients' results came back with some form of co-infection, usually staph aureus. This is a common respiratory staph infection that we can carry but which usually goes unnoticed. However, staph aureus can turn into something nasty if the body doesn't clear it on its own. Some practitioners like to treat staph aureus, and others don't.

In other cases, patients' results came back with a bacterial infection on top of Covid, such as H-Flu (Hemophilus influenza)—which is not your typical flu—or Moraxella Catarrhalis. I have even seen patients test positive for multiple viruses at the same time as COVID-19, such as for influenza A or RSV. The point is, and I'll reiterate this, *we must know what we are treating instead of working blindly in the dark*. Proper testing ensures we can treat sicknesses properly, and only a competent practitioner who understands testing should be conducting them.

The third caveat to understanding COVID-19 is comorbidity. Many people who wear masks tell me they have an "underlying condition," to which I must ask, "Is it stupidity?" Wearing a mask isn't going to help you. Nevertheless, people with chronic illnesses had to be more careful when it came to COVID-19 or any other respiratory pathogen. If you are already sick, diabetic, overweight, or elderly, COVID-19—or any respiratory pathogen, for that matter—can bring you down just as a cold or the flu could. Covid can be brutal because of the transmissibility of the virus and what it can do to your system if you are already run down. If your immune system is weak to begin with, you need to avoid getting sick, period.

When it comes to testing, you can make much more informed decisions about your treatment if you have concrete information. I know I said it before, but we are dealing with the most "deadly virus" known to man, right? You would think people took it seriously and at least made sure they knew exactly what they had contracted when they got sick. When mass testing became available, there were drive-through sites set up all over my state and around the country. People lined up for miles to sit in their cars, get a test, and wait those precious days before their results came in. Originally, you couldn't get a test unless you had symptoms. That was until the government changed the rules and introduced the CARES Act. Now, anyone could get a test at no charge; or at least that's what they told you. The CARES Act had provisions for providers to get paid until they ran out of money. However, the government made it so difficult to get reimbursed that a lot of practitioners and labs charged whatever we wanted for rapid tests because we weren't going to risk getting stiffed by the United States Government on PCR testing. My company never got reimbursed a dime for anyone who was on Medicaid, For those tests that we ran for the uninsured, covered by the CARES Act. We ran thousands of these, and our total payment received was less than ten thousand dollars. Great job, Congress; you screwed that one up, too!

I conducted many tests on people over such a long period of time. Only through mass testing and the recording of data, was I was able to put a lot of things into perspective. For instance, most people recovered quite nicely from COVID-19 at home with no medical intervention. However, there were times when a patient would tell me that they felt like they'd been hit by a truck. I adopted this line and started using it in my daily calls to warn positive patients that their Covid symptoms could flare up and they could be in for something worse before they recover. Although symptoms rarely worsened to that extent, the friend I mentioned earlier, Mike, whose father passed while on a ventilator in hospital, had a brother, Davide, who experienced precisely what I had warned him about.

One day, Mike's brother Davide called me because he wasn't feeling well. As a concerned friend, I conducted a home visit to test him and later called in his positive result. Initially, he told me that he was feeling fine, but a day later, he called me back and said he felt like he had been hit by a truck, exactly as I had warned him he might.

Davide is a relatively young guy in his early forties who works out like a fiend. If any one single person was the definition of "in shape," it would be Davide. However, COVID-19 can hit anyone hard, and some people experience several days of misery. After losing his dad, Davide was understandably anxious about what he had in store for himself, but I reassured him that he would be OK in a few days once his fever subsided. During this period, I had similar conversations with other patients who were flat-out terrified and unsure of what to do. Those who got sick always wanted to know if they should go to the hospital. Unless they were having a medical emergency or were experiencing trouble breathing, that answer was *no*. The hospital probably killed more people than it saved, which we'll cover later when we discuss treatment.

Although I try to educate my patients on how testing works, I'm surprised by someone in my office every day. As of writing this, it is three years after Covid's onset. There is three years' worth of information floating around the world about Covid and testing, and people still don't get it. I had a patient last week who told me they were on Paxlovid. That's the antiviral medication made by Pfizer. When Paxlovid entered the market, doctors on television said it would be a "game changer." My practice prescribed it often during those first couple weeks after Paxlovid entered the market, but most of our patients experienced a serious COVID-19 rebound effect right after their symptoms subsided. Most got even worse on the rebound, and it wasn't pretty. We stopped prescribing Paxlovid shortly after witnessing multiple patients have this effect.

So, when this patient told me she was on Paxlovid, I asked her who was prescribing it. She told me that she got it from the hospital after she was diagnosed with Covid. I then asked her *how* she was diagnosed. "On a home test," she said. Insert eye roll and hands being thrown up in the air. *Who in the world* would prescribe this antiviral without the patient being properly tested and diagnosed? Again, it is for reasons like these that much of the medical community has lost credibility for the way they handled Covid.

Testing goes a long way when you have accurate information. If we don't know what we are dealing with in totality, we are just guessing, and that, my friends, is not the way we should be handling something so serious. Most of this blame goes on the patients, though. Their eagerness to get back to work, get around the results, and look for the best scenario regardless of accuracy, really is a letdown. My team and I worked our asses off to make sure we were there for the public right from the start, but it was often unreciprocated. I completely understand why patients were so frustrated because I believe the government overreacted when enacting all the lockdowns and ridiculous criteria, but if you're going to push back on the science of testing, why come to me for a test in the first place?

To understand why I believe the federal government botched this whole thing, you must understand reporting to keep up with data analytics. When one of our labs reports a positive case, they must report it to a state agency. If they are not set up for reporting in a certain state, the onus then falls on my staff and me to report the numbers. My team had another layer of nonsense put on us when we started conducting rapid tests, as we'd have to call or email the list over to the county agency in charge of communicable diseases. This way, the county agency could conduct contact tracing. That one was fun, right? However, once home tests were available and people took matters into their own hands, none of those home positives got reported. Most people, not all, did *not* run to an office like mine for a confirmatory test. Most people ignored the results completely or weren't aware of any co-infection they may have had. They had no way of verifying the negative results, considering you need a huge amount of virus to turn that rapid test positive.

My point is: those case numbers you saw right in the beginning of the pandemic were just our ability to catch up to positives with the number of tests out there. The numbers reflected a few months into the pandemic are fairly accurate because, during most of 2020, people were trampling over each other to get a test, and mostly only PCR testing was available. After the advent of the rapid test and rapid home test, the reported numbers you saw weren't even close to what was floating around. For example, if you saw a thousand reported cases, that's probably only a quarter of the actual cases when you realize that only a small fraction of the population runs to a medical office or hospital anymore. Only those tests conducted in an office or hospital are reported. The rest are up to the individual at home, and you will never hear about them!

The federal government took away any semblance of seriousness once they started shipping free home tests to people's houses. Knowing what we know now about the accuracy and the amount of virus it takes to turn a rapid home test positive, it's obvious that the government didn't really care about tracking, eliminating, or planning for Covid. I was suspicious once I saw early data and figured the government narrative would catch up with me soon. Why would the federal government disrupt people's lives, livelihoods, and schedules once they knew the truth? Well, it seems it was all for money and power! When my friend Adam asked me when I thought we would be back to "normal" in March 2020, I remember telling him that once we delivered the data we saw on the ground, the pandemic should be over by September. He was shocked at that prediction and was frantically trying to plan ways to keep his business alive and, more importantly, feed his family. My six-month prediction was unsettling for him, to say the least. Who knew it would drag on well beyond that?

I felt really bad for Adam when the pandemic continued dragging on, but he adapted and overcame. I can't say the same for some others, though. I saw restaurants shut down and mom-and-pop shops go out of business after many years as a community staple. I saw people lose their jobs and a lot of companies cash in on free Covid money. This was one thing my team and I never participated in. First, we were too busy. Second, it wasn't the right thing to do. In hindsight, and after we got screwed out of all that CARES Act money for testing, I kind of wish we had applied. We conducted a LOT of tests for free and, just to note, my staff don't work for free!

CHAPTER 10

Antibodies and Antibody Tests

This book wouldn't be complete without talking about antibodies and antibody testing. While virtually nobody talks about this now, it was a hot topic in late 2019 and throughout all of 2020. I don't think most people really understand how this works, but our company got some very good data on antibodies during our testing cycles. Patients would call and ask if we had antibody testing because they thought that if antibodies were detected, it could show a previous infection and mean they were safe from catching Covid moving forward. Think about that for a second; a test to see if you already had Covid or, as most would say, "a virus so deadly, you have to test to see if you have had it."

I've said this before, and I'll say it again, never in the history of the world were asymptomatic people—those with ZERO symptoms—begging to get tested for *anything*. Sick people go to the hospital; sick people go to the doctor's office. In this bizarre universe, perfectly healthy people were begging to see whether they had this virus that was causing all this destruction. Most people assumed that if they had antibodies, they had the golden ticket. Too bad antibodies don't last forever! The opposite argument was made for people who thought they dodged a bullet; they thought they were invincible or that the cloth mask they had on was providing them with a Covid force field. They would say, "I had my antibodies tested, and I was negative, so I never got it!" All these people were dead wrong!

First, we must understand antibodies and how they work. Antibody testing is conducted using blood that is sent to the lab for testing. Yes, there were finger-prick tests on the market that claimed to test for these antibodies, but they are highly inaccurate, so I won't spend any time on them here. Proper tests are sent to a reputable laboratory that looks for a few things. Typically, a lab will test for two to three types of antibodies: IGM, IGA, and IgG. If a sample comes back reactive for IGA or IGM, this is indicative of a current infection. Most IGA or IGM antibodies will start to tail off after about two weeks. IgG antibodies are your longer-lasting antibodies. These can last from weeks to months and are dependent on the patient, the severity of the infection, and an abundance of other factors. Understanding this information is crucial to discussing vaccines in chapter 12.

Some of our regular clients asked about conducting weekly antibody tests in addition to our weekly COVID-19 swabs. This scenario was impractical, but our clients were just like everybody else in the world:

scared, proactive, and grasping at straws. They would throw whatever they could at the wall just to see what stuck. I had multiple conversations with police chiefs, city administrators, and healthcare professionals as to the pros and cons of doing these tests, but it all came down to data, as this would be great information to have.

We had months' worth of data on thousands of patients, which gave us a timeline. This initial testing data demonstrated who had been infected, how many of those positive cases developed symptoms, what those symptoms were, and when and if the patient got reinfected. A lot of administrators initially asked for negative tests so their employees could go back to work, but this was wildly inappropriate. I would ask them, "Do you go back to the doctor when you have the flu to see if you are over it?" The answer was an obvious no, but COVID-19 scared the shit out of people, and many didn't want to assume the liability. I would get very upset if an employee called me, telling me that they "needed" a negative test to go back to work. "Who is your supervisor?" I would say. Then I would make a call and relay that information to said supervisor so I wouldn't get a hundred more of these needless requests.

When someone comes back positive on a PCR test, they can keep testing positive for up to ninety days while they shed dead cells. The longest I have personally seen in one of our patients was forty-six days. This was very early during the pandemic when patients wanted to donate plasma to hospitals. The hospital kept turning her away because she was still testing positive. However, by the time she finally tested negative, her antibodies were too low to donate. If a patient keeps testing positive, it does not mean that they are contagious. If you remember timelines, they are only contagious for up to fourteen days.

It is inappropriate to force someone to do something that means nothing scientifically and disenfranchises them from earning a living. However, once COVID-19 came around, the entire world started making up its own medical rules, and most of these rules weren't practical. Essentially, the inmates were running the asylum, and a lot of unwilling participants followed suit because there was nobody manning the guard house.

When we decided to start conducting antibody testing, we learned a lot. First off, if we were going to do this, we were going to have to commit to doing it over a period so that we could see a timeline. We would be looking at how long the antibodies lasted and which kinds of antibodies they were. We would have to back that data up with PCR testing on the same day we conducted the antibody tests. All these patients were essentially participating in a scientific study.

If someone were to come back "reactive" for antibodies, either IGM, IGA, or IgG, we would need to confirm that result with a PCR test. In other words, an IGA or IGM "reactive" would be indicative of a current infection. We needed to verify that with a PCR test. A "reactive" positive for IgG would indicate a previous infection, and our PCR could loosely verify that the infection had passed. If the PCR came back negative, we could assure the IgG was in the past, considering that a person can continue to shed inactive virus cells for up to ninety days. We could also verify this with previous PCR testing considering that we were testing these patients on a weekly basis.

So, if patient John Doe was positive on March 1st and continued to test positive throughout April, we would know that he was not contagious after March 15th or thereabouts. However, we should observe antibodies that were reactive accordingly. If his IgG was present in May, we would have a timeline congruent with a negative PCR in May. I know this sounds like a lot of science, but honestly, it was just common sense. We had a patient population that was willing to participate, and testing this population throughout the pandemic could teach us a lot. When we conducted those antibody tests, all the participants who came back "reactive" for IGM or IGA also had a positive PCR test. For most patients who had a "reactive" IgG antibody result, their current PCR test was negative; however, we could look back at their testing results from previous weeks to determine when they had tested positive.

What does this all mean? The bottom line is the findings showed that patients who had antibodies from a previous infection had an average of 120 days of natural immunity. This duration, however, does NOT take into account natural immunity from B and T lymphocytes, which is more difficult to test for and not widely available. While some patients had reactive antibodies for a few weeks, one individual maintained "reactive" antibodies for an entire year. Roughly fifty patients participated in the antibody study, which was conducted several times throughout the pandemic. One of the most interesting discoveries made through this testing was when we tested the same group of patients AFTER they received vaccinations. The timeline was the same! All vaccinated patients showed "reactive" antibodies, on average, for four months after the vaccination. Although the vaccines generated reactive antibodies, these effects were short-lived and did not provide any long-lasting or permanent protection against COVID-19.

Patients initially assumed that we could test them to see if they had EVER had COVID-19. However, if they had been infected last year or the year before, those antibodies were long gone. Furthermore, having a positive result does not necessarily mean that a person developed symptoms as most never did. To definitively confirm that you have NEVER had COVID-19, we would need to have conducted weekly PCR testing on you for the past three years. Luckily, we had the luxury of doing so, which is why we obtained such great early data that helped us put COVID-19 into perspective and understand its totality. The metadata we collected on antibodies was invaluable because it coincided with the data we had from PCR testing. As of the time of writing, three years after the start of testing, I rarely hear the word "antibodies" except to reinforce how it contributes to our overall understanding of COVID-19.

CHAPTER 11

Treatments and Natural Immunity

The practice of medicine is just that, a practice. I employ multiple practitioners, and I would never tell them how to practice individually. However, they all share information and evaluate what may work. Our company protocols regarding Covid remained mostly the same throughout the pandemic. They included taking vitamin C, a double dose of vitamin D, Zinc, and fluids with electrolytes. We also recommended Tylenol for any aches, pains, or fever and lots of rest. We initially thought that ibuprofen flared Covid up but later learned that it was safe, so after a while, we recommended rotating between Advil and Tylenol. The max dose for Tylenol is 3000mg per day, and some people were getting a little Tylenol-happy which could do a number on your kidneys.

We also learned that flu protocols flared COVID-19 up. Things like Tamiflu and Elderberry had a negative effect on those trying to recover. Yes, you heard that right,

Elderberry! Who would have thought? I give practitioners a lot of credit here. Sharing information is crucial when dealing with a novel virus. The world had never seen Covid and had never seen anything on this scale that crossed continents so quickly. We must go all the way back to the 1918 Spanish flu to make any comparison that resembles this outbreak. In the last hundred years, we've seen plenty of pathogens that have set alarm bells ringing but nothing that showed up on every person in the world's backdoor within months.

A lot of my friends and colleagues were working on the front lines in the hospitals. They had to deal with atrocious conditions like wearing an abundance of PPE and masks for twelve-plus hour shifts. They put more people in body bags than they would care to remember, and the trauma from that alone would give the strongest of caregivers a ration of lifelong PTSD. I heard stories of perfectly healthy individuals, some younger than you might think in their early forties, diagnosed and doing well in a hospital bed, only to spike a fever and be placed on a ventilator three hours later. It was mass chaos. Patients would flood the hospital out of fear and take up valuable resources when they may have recovered just fine at home.

One of the biggest things we learned during the pandemic is a process called "proning." You see, your lungs are located more toward your back. Therefore, people's lungs would fill with fluid and limit oxygen exchange because most people lie down on their backs, especially if in a hospital bed. Prone positioning beds or Rotoprone beds have been available since the early

2000s and have been used for acute respiratory distress syndrome and many other types of severe respiratory failure. These beds are very large and very expensive. They, for lack of other words, rotate the patient like a rotisserie chicken. Their heads and body are secured, and the patient can be placed on a ventilator as the lines come up through a hole near the head. The availability of these beds was scarce as so many patients had been placed on vents in the ICU.

In lieu of a prone positioning bed being available, hospitals sent out "prone teams" to manually visit each patient and prone them into a face-down position. This process would continue until every patient was turned and then repeated every few hours. Proning reduces the mortality rate for patients in acute respiratory distress by a large margin. This allows for better oxygen flow to the lungs and blood flow through the vessels. However, proning can take multiple staff members to accomplish because it involves more than just turning the patient. It encompasses ensuring the patient's lines are correct, the patient is comfortable, and all medications and hospital equipment are attached correctly and functioning. Another important factor in this process was pillow placement. Most patients on vents are heavily sedated and not moving on their own. This can cause tissue breakdown and the start of wound formation. The placement of the patient was of utmost importance because it could lead to additional complications on top of what they were already being treated for.

I would often have patients who were concerned about their blood oxygen levels call the office. We were having a hard enough time getting our hands on medical equipment, and part of the reasoning behind that was that everybody in the world was buying up all the equipment. One of these pieces of equipment was a pulse oximeter. Those are the tiny electronic devices that clip onto your finger to measure blood oxygen levels and heart rate.

One patient called and told me that his pulse oxygen was at 88 percent, but he was not having trouble breathing. When we told him to lay on his stomach, take a nap, and call back in thirty minutes, he was confused. Nevertheless, he did so, and when he called back an hour later, he said, "Holy shit, I'm at ninety-five; how did you do that?" This process worked well and probably prevented him from occupying a hospital bed. It also showed him that he could manage Covid with some easy home remedies that could make him feel tremendously better. This gentleman recovered quite nicely at home, as did most of our patients.

One of the reasons that so many people died early during the pandemic was the lack of knowledge we had regarding this disease. The medical community saved a lot of lives, but some of the early practices we used, unfortunately, took more lives than I would care to think about. Patients were placed on ventilators *way* too early! The typical standard of care is to place someone on supplemental oxygen when their pulse oxygen level reaches 91 percent, or the patient is showing signs of distress. In cases where an arterial blood gas level was taken, patients with high CO_2 levels were intubated immediately as their bodies weren't able to compensate.

However, this approach caused many patients to reject the ventilator therapy, which in turn meant they required more medications and heavier sedation, further prolonging their time on the ventilator and lengthening their hospital stays.

This early treatment process killed the patient in more cases than fathomable. If prone positioning beds had been used at this earlier stage, more of these patients would have survived or experienced less of COVID-19's long-lasting effects. Now that we've gone on our learning curve to treat Covid, the protocol is to use high flow oxygen, which is different from supplemental oxygen. This process includes using warm, humidified gas to reduce airway dehydration and increase secretion clearing.

Sometimes, patients showed Ground Glass Opacities in their lungs following an x-ray. This is a whited-out chest x-ray showing fluid in the lungs, which is essentially pneumonia or Covid pneumonia. Hospital treatment for this could include high flow oxygen, IV steroids, or a ventilator. One of our local hospitals has thirty ICU beds, and in early 2020, all of these were full of ventilated COVID-19 patients. Then, more floors at the hospital were commandeered to accommodate the overflow of COVID-19 cases. About 95 percent of the patients being treated at the facility were COVID-19-related. In comparison, at the time of writing, that same hospital has no more than five intubated patients at a time. Less than half of one floor is isolated for COVID-19. This reduction is due to understanding how to treat Covid as a result of facilities sharing information.

Furthermore, the death count, which was revised each time I saw a new news report, was blatantly overinflated. Many people would need hospital treatment for any number of accidents or diseases. Many of those would go into those same hospitals without having contracted a respiratory disease but acquire it during their hospital stay. COVID-19 was simply EVERYWHERE during late 2019 and early 2020. When someone died of any number of instances but was also positive for COVID-19, the blame was placed on Covid. Don't get me wrong; Covid is nasty. It's as nasty as influenza A and is even more transmissible.

Transmissibility is one of the largest factors concerning COVID-19; it can spread so rapidly and easily from one patient to another. The R Naught is the basic reproduction ratio or rate used to measure the transmissibility of an infectious agent. The initial R Naught of COVID-19, as measured by the WHO, was between 1.4 and 2.4; however, additional studies have shown that the R Naught for COVID-19 can range from 1.5 to 6.68. What does this mean? Essentially, for every person that contracts the virus, they can spread it to roughly three to four people on average, and in some instances, even more. Ever heard of a superspreader event?

Since everyone had COVID-19 in late 2019 and early 2020, most people were dying "with" Covid and not "from" Covid. Yes, many people succumbed to this disease early on during the initial outbreak, but this was for various reasons. We've already discussed the problem with hospitals and our medical community trying to share information and find appropriate treatment regimens.

There was also the fact that this virus was novel! I would tell patients that if they are frail and susceptible to dying in the next five years, a cold could bring them down, the flu could bring them down, and Covid could bring them down. When the pilgrims landed in the New World in the early 1600s, many of the indigenous people got sick and died from rare diseases their bodies had never seen. While the pilgrims settled new lands, they had no idea that they were bringing disease and destruction along with them to humans who had never been exposed to things like smallpox!

Let's talk a little about monoclonal antibodies. Remdesivir was the first approved monoclonal antibody treatment for COVID-19. It was approved by the FDA in October of 2020. This is one of many antibody treatments like Regeneron, Actemra, and Bamlanivimab, to name a few. Remdesivir was most widely used in my area, but some of these have been discontinued and had FDA approval revoked because they are ineffective against certain variants of COVID-19. We didn't have much experience with monoclonal antibodies in our offices except for the instances where we sent patients to antibody clinics available in Florida that Governor Ron DeSantis had set up. Occasionally, a patient would ask about them, and our thought was, *nothing could really hurt*. Whether they worked or not was another story.

In the hospital, Remdesivir was used when a patient didn't respond to Tylenol given for a high fever and if they were on supplemental oxygen. This was also dependent on the patient's comorbidity, as it would also be used in cases where someone was diabetic and could not get their sugar levels under control. Normal protocol in our local hospital was to prescribe six doses of Remdesivir to patients, but in a lot of cases, those patients would improve after three treatments and be discharged before needing all six. Was this because the Remdesivir was working, or was the patient just running the course of the virus? Science says that Remdesivir stops the virus's replication in the body. However, in the four thousand positive cases I saw in my patients, the virus ran its course naturally with simple over-the-counter treatments.

Paxlovid was introduced in December 2021 and was the first oral medication authorized by the FDA for the treatment of COVID-19. Paxlovid is an antiviral which stops the replication of the virus in the human body, much like Tamiflu does for influenza. It consists of two separate medications packaged together. This medication inhibits a key enzyme required by the Covid virus to make functional virus particles. Paxlovid medication needs to be taken within the first five days of onset; otherwise, it will have no effect and works much the same as antiviral medications like Tamiflu for influenza.

Once approved, our office started prescribing Paxlovid for COVID-19 positive patients who were presenting with symptoms; however, most patients started getting a nasty rebound effect. Their symptoms would start improving, and then, seemingly out of nowhere, they would develop more severe and debilitating symptoms. One of my practitioners began prescribing two courses of the medication once this started happening, which led to fights with the pharmacies she had called the orders into. As of today, we don't prescribe this medication anymore unless a patient demands it, and even in that case, we will still push back and try to educate them. It wasn't too long ago that Dr. Anthony Fauci caught Covid, and he himself started taking Paxlovid. When he went public with his treatment, we all sat back in my office and just waited. In a matter of days, he had the same rebound effect. Guess he should have called our office first!

This chapter wouldn't be complete without mentioning Ivermectin. We were not using this in our offices, but I worked with a lot of doctors who were! Ivermectin is an antiparasitic drug that was first discovered in 1975. Its initial indications were in veterinary medicine for the treatment of heartworms. However, it was approved for human use in 1987 for a variety of indications. Ivermectin has been debated profusely over and over across many platforms on the COVID-19 stage. I can't stand it when someone misleads the public regarding such a medication . . . though I may be about to go down a different road than you think.

We had no idea what would work during the initial stages of the pandemic, and doctors have a right to try everything to see what makes a difference in someone's treatment. Ivermectin is an antiparasitic drug, and COVID-19 is a virus. However, a virus is classed as a "small obligate intracellular parasite." Therefore, a virus is "like a parasite," so it makes sense for doctors to try antiparasitic medications to see whether they can alleviate a patient's symptoms and potentially save their life.

When I heard someone on TV say doctors were irresponsible for giving patients "horse medication," I simply regarded their comment as horse *shit*! The people throwing out these comments are clearly not the experts; they probably never even took five seconds to research the medicine and see that it was approved for human use in the '80s. They called these doctors irresponsible, but they were just trying to further a narrative which is one of the most blatantly irresponsible things I have ever witnessed. Did the medication work? Well, that's another point of contention. None of our practitioners were prescribing it in our offices, although we did have multiple patients ask for it. There was a doctor I consulted with in Key West who claimed a lot of success with the medication, so if we had a patient inquire, we would usually refer them over to his office.

Sometimes, the best answer any doctor can give a person is "I don't know." This rarely happens today due to the fear that they may lose a patient to another practitioner, but the reality is nobody knows everything about everything. I can tell you that I do not have enough data on Ivermectin because my team didn't use it, but I can give you an opinion that may help. We used some of the most minimalistic treatment methods for patients, and in my experience, most patients didn't need any medical intervention.

If a patient spiked a fever, they'd be on the road to recovery once their fever broke after a day or two, and they would be back to baseline health in two or three days. When observing the patients that used Ivermectin, the result was the same. I would be curious to put Ivermectin patients up against those with a placebo to see if it really shortened the duration. I would think the results would be the same, but at the end of the day, it couldn't hurt and was worth a shot. My hat is off to those doctors who forged new paths to try to help the world turn the corner!

This brings us to natural immunity. When we first started working with COVID-19 patients, the big question was: how long does natural immunity last? When I got sick in late 2019, we were not yet testing for COVID-19, so I could not verify it. It wasn't until I started testing people in their homes and recognizing the symptoms and signs that I fully understood Covid was what I'd had. I'd lost my sense of taste and smell and had a dry cough that lasted for weeks and weeks. I'd been exhausted, had headaches, and basically felt like a truck had hit me. On top of all that, I knew that if one person in a household contracted Covid, everyone in the household had it, but I had been in and out of homes, one after another, and wasn't catching Covid. I always say that I hate to use myself as a data point and that one person is not a data set, but . . . when and if I were to catch Covid again, I would acquire at least one point of data on natural immunity.

During the pandemic, we found we could compare this coronavirus to other coronaviruses and respiratory pathogens. However, I'm sure you have heard or even been a victim yourself of catching the flu twice in one year. The longer a virus propagates, the more chance there is it will mutate into various other potential infectious variants.

To gather real data on natural immunity, we looked at my own patient population. Although we had thousands of patients and over four thousand positives in total, at the time we were discussing natural immunity, we had about a thousand positive patients to reference data from. At first, we had no reinfections, which seemed like promising data regarding natural immunity. However, at the eighteenth-month mark, in September 2021, we had our first reinfection. Then, like clockwork, the numbers started to increase. First, we had one, then five, then twenty, and finally, I caught Covid-19 again in October 2021, twenty months after my initial infection. While this data allowed us to start predicting susceptibility and observing seasonality, it was not a straightforward solution.

Two months after my patients started becoming reinfected, in December 2021, there was another round of Covid that rushed around the world. This was called the Omicron strain, and those with the reinfections became reinfected *again* a short two months later. It wasn't just *some* patients either. It was an abundance of them! I was also included in that mix. The second time I caught Covid-19 in October 2021, I felt awful and experienced lots of aches and pains. I'd worked out my arms at the gym and ridden on my jet ski a day before the onset of symptoms, which amplified the muscle pain Covid can cause.

I sat in bed for days and could barely move. I tried to stay hydrated but couldn't focus much as my attention was on the pain in my arms and quads. I lay there wishing the pain would subside, but it continued for days and days. Compared to this, the symptoms I experienced in December 2021 were mild at best. I'd felt like I had a slight cold coming on, so I ran a full panel, thinking that there was no way I could have had Covid-19 again two months after my last infection. Nevertheless, the doctor at the lab called me personally and said, "You really need to be careful; this is your third time with it." We both laughed it off because we had become experts with Covid, and neither of us was worried. I overcame that bout, too, in a few days, and my mild symptoms subsided very quickly.

The response of most of my patients to the Omicron strain was the same, and there are a few reasons for this. First, when your body begins a natural immunity response, it targets all the proteins inside the envelope of the virus. Your body builds B and T cells, which give you long lasting immunity. The B cells being your memory cells and your T cells being your helper cells. On the other hand, the spike protein is what mutates on the outside of the virus, tricking your body so that it can bind to your cells. However, once it does bind, your body reacts to the spike protein. This means that your body has built natural immunity to many of the proteins within the envelope, which is why you may experience a milder reaction each new time you get infected with Covid, provided it has a chance to replicate in your system.

From this, we could also see that the virus had been mutating for quite a while and that people can become infected or reinfected as multiple strains are present at any given time. If you were to make a copy of a copy and then copy *that* copy and so on, you would wind up with a copy that doesn't resemble the original at all. That is what happened with most copies of the virus. There are thousands of versions of Covid-19 or at least the spike protein, but those variants eventually burn themselves out, which is why you never hear about them. The ones you hear about are the ones that learn how to bind and become the most prevalent among infected individuals.

There are varying reasons why some patients with Covid experience more severe symptoms than others. However, now we are seeing versions of the spike protein that are weaker in comparison to others, which could be a contributing factor to the lesser reaction. Another reason could be due to the timing between positive cases, as antibodies typically last an average of 120 days. We began to question whether a combination of antibodies and memory cells was coming into play during Omicron, but the spike protein looked completely different, causing the body to be fooled and the patient to develop reinfection. There is still a lot to learn about COVID-19, but we knew more about this thing earlier than the media led and CDC would have you believe.

As we continue to study COVID-19, treatments are continually changing. However, I said something in March 2020 that seems to have come full circle at the time of writing this in 2023. I noted that we have quite a few things on our RPP, like the flu and H1N1, and that in a few years, Covid will just be one more thing we test for on our panel. Nobody would care about it, but at least we'd know how to treat it by then. It seems like we got there relatively quickly, although with so many patients and so much practice available, reaching this point was and is inevitable.

CHAPTER 12

Vaccines?

Despite all the science behind viruses, testing, treatment, etc., the most common question I get every day is regarding vaccines. Usually, the patient asks me whether they can pick my brains, and I always wonder which direction this will go in. Nevertheless, I always hear the exact same line next: *How do you feel about the vaccines?* I usually start by explaining what a virus is and how it transmits so the patient has a better understanding of how effective or ineffective vaccines can be. I wrote this book because I got tired of explaining this to patients one at a time! It would be great if those patients spread the message, and then those they told did too, but we aren't starting an Amway franchise, nor is the science behind vaccines that easy to understand.

First, an introduction to vaccines. It's important to note that scientists have not been working with viruses and vaccines for all that long. The first vaccination in recorded history was created in 1796 when Dr. Edward Jenner successfully vaccinated an eight-year-old patient using puss from a Cowpox sore after discovering that people with Cowpox were immune to smallpox. Scientists didn't realize that these diseases were viruses because they didn't have the equipment able to see something so small. They just knew that they were "nonbacterial" illnesses. The first virus ever discovered was the Tobacco Mosaic virus in 1930. While Tobacco Mosaic has been known since the 1800s, it wasn't determined to be a virus until only eighty-nine years before COVID-19 made its appearance.

Up until the inception of COVID-19, the vaccines we are accustomed to are called "live-attenuated vaccines." These vaccines use a weakened version of a virus to help people build immunity without experiencing any symptoms of the original virus. The first attenuated vaccine came in 1885 from Louis Pasteur. He attenuated the rabies vaccine in rabbits and harvested it from their spinal cords. Sounds gross, right? It wasn't until after the 1930s that attenuated vaccines became commonplace. After this point, there was mass production of vaccines for yellow fever, pertussis, influenza, and polio.

A lot of people throw around the word "vaccine" without really understanding the nature of what vaccines are, how they work, or their effectiveness. The old definition of the word vaccine stated it was a "product that stimulates a person's immune system to produce immunity to a specific disease." However, the CDC changed that definition in 2021 to a "preparation that is used to stimulate the body's immune response against diseases." This change was a little disturbing considering most of the populus was used to comfortably and clearly understanding terms and their definitions.

Adults who have completed their series of polio vaccinations are considered immune for life. For people who have been vaccinated with an MMR vaccine, they are considered immune for life as well. A smallpox vaccination provides full immunity for three to five years before it starts decreasing in strength. However, the smallpox vaccine has shown to be effective in 95 percent of those vaccinated. A flu shot, which is now considered a vaccine, has an efficacy rate of anywhere from 30 to 60 percent depending on the study you read and the year in which the vaccine was administered. It's no wonder why the flu shot has always been referred to as a "shot" instead of a "flu vaccine." Nevertheless, flu shots are recommended during flu season, and it's common knowledge for those in regular receipt of flu shots that they should get one yearly.

Why is this important? I think it's obvious, considering respiratory viruses affects the older population even more than the young. Most of the older population's knowledge of how vaccines work is cemented on the common vaccines we mentioned earlier. Most of society does not have a degree in virology or epidemiology. They don't have the luxury of calling an immunologist daily like I do to discuss different hypotheses regarding COVID-19. The media did a horrible job at relaying this information, and most "experts" just regurgitated what other TV "experts" were saying. Most just repeated CDC talking points. I watched like the rest of you and never heard anyone cited who had tested and treated thousands of COVID-19 patients. Their opinions were never based on data analysis but rather on their own predictions and blatant conjecture.

We had real data as early as April 2020 and were very vocal about our findings to our patient base. However, the things we were saying in our office didn't make the news cycle for twelve to eighteen months in most instances. Sometimes even longer! The information was there, but it seemed our data didn't fit the narrative of the day, so it was either looked at with wide eyes or completely dismissed until the reality of infections, transmission, and vaccines began matching our data and predictions.

Once the vaccines were released, we told our patients that we lacked data on them because the Pfizer and Moderna vaccines utilized mRNA, a completely new technology being used on a mass scale. The Johnson and Johnson (J&J) vaccines were "Viral Vector Vaccinations," which is more a traditional technology. We told patients that you could still catch and transmit COVID-19 even when vaccinated and that they would most likely need a vaccine every year, like a flu shot, maybe more depending on the vaccine's efficacy. Meanwhile, President Biden was direct and center on television, calling this a "pandemic of the unvaccinated" and pleading with people to get their shots. President Biden said, "if you get the vaccine, you can't get Covid and you can't give Covid to someone else". As we first watched this, my staff looked at each other, and an audible groan went around the room. "How is he saying this?" someone said. "He's blatantly lying, or someone is lying to him, and he doesn't have a clue," someone else replied.

As a business, we had to transition out of testing and into vaccines. People were starting to focus less on testing and more on getting their vaccinations, and we wanted to be ready and available to handle the volume. In the end, we never did administer a COVID-19 vaccination in our New Jersey office, but we did have J&J and Moderna readily available once we were able to get ahold of them in Florida. The mRNA vaccines were touted as having a roughly 95 percent efficacy rate after two doses. The J&J vaccine boasted a 77 percent efficacy rate, but the patient would only need one shot—they were one and done, as everyone liked to say.

The vaccines came out a few months before we could secure them in the office, and we noticed a few things. When COVID-19 started, I rolled my eyes every time I called a positive case who was stunned by their diagnosis. *But I wore my mask the entire time*, they would say. Once the vaccines were released, the by-line changed. Then when I called a positive case, they said, *But I was fully vaccinated!* Patients were confused, as the media and the US government were telling them they couldn't catch COVID-19 if they were vaccinated, but there they were, sick with COVID-19! I'm a data guy, and I love putting two and two together. Therefore, we started recording the data on vaccines and positive infections a few months into the mass vaccination program. While doing so, we discovered it was worthless to record this data as the trend showed there were no clear distinctions between any single vaccine and positive Covid cases, and there weren't any specific vaccinated patient groups catching Covid more than others.

My team and I were additionally asked to conduct more antibody tests because some of our clients wanted to know whether the vaccines were causing an antibody response and, if so, for how long this response lasted after their employees were administered a vaccine. *What we found was astounding!* Most patients we called to present positive results were vaccinated roughly four months ago or more. This matched the antibody testing results. There was an average of a 120-day or four-month antibody response from the vaccinations before the IgG levels tailed off. Some patients responded for longer durations, and some shorter. The immune system response in vaccinated people versus the response from those who gained natural immunity following a Covid infection was the same except for one BIG difference!

Vaccinated people did not receive the same B and T lymphocyte response (long-term immunity) as patients who acquired immunity following a natural infection. Essentially, these vaccines were acting exactly the way flu shots did. It's hard to gauge the effectiveness of the vaccines in conjunction with what the pharmaceutical companies were touting. When Pfizer said they were going to be close to 95 percent, did they mean for four months or for life? The older the population who were used to specific language thought this percentage meant the latter!

There is a scientific reason why patients were not receiving the same immune response from the vaccine as they did from natural immunity, but we needed some data to back it up. A virus contains roughly twenty-seven proteins; the spike protein is what mutates, but the inside of the virus stays the same. It turned out that the vaccines gave people a reaction to the *spike protein* and NOT the entirety of the virus. In the case of J&J, for example, this vaccine uses an inactive cold virus to carry genetic material, prompting your cells to create a harmless piece of the spike protein. mRNA vaccines use mRNA (messenger RNA) created in a laboratory to teach your cells how to make a protein or a piece of protein. This triggers an immune response in your body and leads it to create antibodies. Simply put, the mRNA instructs your body to make a harmless version of the "spike protein."

The key phrase I want to emphasize here is "spike protein." While antibodies are typically borne from the spike protein, the B and T memory cells are rooted in the entirety of the virus. This means that nobody was developing long-term immunity after having these vaccines, and it was blatant malpractice and a bald-faced lie to tell people that they could not give COVID-19 to someone else once they had been vaccinated. Just as some people thought a cloth mask created an anti-Covid force field, they also thought these shots would make them invincible to the virus.

Furthermore, we began seeing some very questionable things in our vaccinated patients. Again, I'm a data guy, so I look at cause and correlation. The top three things we mostly saw after someone had an adverse reaction to the vaccines were heart issues, joint issues, or clotting. With Pfizer, it was mostly heart and joint issues. With Moderna, it was usually joint issues. If a patient had clotting, it was a toss-up between Pfizer, Modera, and J&J. Yes, I know there are other vaccines on the market, but these are the three that were most prevalent and the ones I have data on. I can't speak to the others because I simply do not know. However, vaccines all work the same way, so you can draw your own conclusions after researching the different types.

One of our patients came to me after getting a series of Covid shots. She was twenty years old and worked as an EMT. She was essentially forced to receive the vaccination series because her department would have fired her if she refused. She had her first Pfizer dose, and within a week, her joints had started swelling up. After receiving her second dose, her joints got worse. They were tight, inflamed, and she was unable to fully grip anything. She came to us looking for help. The first thing I said to her was, "So, you had an mRNA vaccine. Either Pfizer or Moderna."

She looked at me, paused, and said, "How did you know that?"

I said, "You're not the first and won't be the last!"

My patient was having some serious joint issues. She was gaining weight and was unable to bend. Her body was turning stiff, and did I mention she was ONLY twenty years old? I feel the need to re-emphasize this because this wonderful kid has her whole life ahead of her. We are only primary care, so we did what we could, but eventually, we referred her to Orthopedics and Rheumatology. She came to us because we were the Covid people, but we weren't equipped to handle some of these serious complications. Eventually, she came back to us because none of those specialists could do anything to help. She was engaged and complained that she could not take her engagement ring off because her fingers were so swollen.

I decided to try something different and asked her to come in for a NAD infusion. I told her that I wasn't sure if it would work, but we were willing to give it a try. Three days after the NAD infusion with a glutathione push, she walked in with a smile and slid her engagement ring on and off her finger. In that instance, NAD worked to repair the mitochondria of her cells. We use this treatment commonly now for serious inflammation. In my patient's case, it did the job, but it was short-lived. She started to swell again after about three weeks and only got about three months of lingering relief from the NAD. It seemed as though she was going to have long-term issues related to her vaccines.

In doing some research, we found that her mother has a rare disease that she developed in her late forties or early fifties called Charcot-Marie-Tooth Syndrome. It is hereditary and is usually developed in childhood or early adulthood. Our patient was most likely to inherit this from her mother, but the timing of her symptoms was alarming. Did the vaccines speed the disease's progression? Cause and correlation would tell you yes. These symptoms developed immediately after her first shot, got worse with the second, and she became horribly inflamed when she caught Covid a few months later. This patient is one of many with joint, heart, or other issues that seemed to appear right after her vaccinations.

In September 2021, we hired a new executive assistant, Caitlin. She honestly didn't have any experience but was eager to work and reminded me a lot of myself. I felt bad for her, as I was jet-setting back and forth between states and other responsibilities, so she was left to her own devices, but she learned quickly. We were already one-and-a-half years into the pandemic at that point, and she was thrown straight into the fire as the team was spread across multiple sites testing people while trying to maintain our office. She learned a lot because she paid attention to detail and asked questions. Once she picked things up, she ran our New Jersey office and became the team's go-to person.

Caitlin had access to our data and spoke with patients often. When her own family was debating the vaccines, Caitlin emphatically implored them NOT to do it, especially the kids. There was no reason for it, as kids were virtually unaffected by the virus, and she saw how ineffective the vaccines were. She has also experienced COVID-19 personally and recovered quite nicely. She had the flu in the same year and, in her words, said, "I'd much rather have Covid. The flu absolutely sucked." However, her family was adamant about getting everyone vaccinated. She told me she was concerned about her cousin. He was thirteen and being treated for atrial fibrillation. It was under control, and he had been seeing a cardiologist regularly. His own cardiologist recommended the vaccine for him! It wasn't long after we had this conversation that she lost the battle with her family. Her cousin got the shot and two weeks later died of a sudden heart attack. You will never see that statistic in the Vaccine Adverse Event Reporting System (VAERS). In fact, you won't see any of the stories I tell you in VAERS.

The reporting system isn't set up the way you think it is. When there was a positive Covid case, either our office or our lab must report that positive to the state and, in turn, the county, etc. Those positive cases coming out of labs and offices were fairly accurate, at least at the beginning of the pandemic. They painted a relatively truthful picture of the number of positive cases at that time and in what area. As we've discussed, once the federal government screwed that up by shipping free, inaccurate rapid home tests to everyone, those numbers got skewed. People who do what I do hate data that is skewed because it becomes unreliable. In this vein, the VAERS numbers do not have to be reported, and if they are, they may be discounted on the grounds that the adverse reaction was too long after a shot was administered. Additionally, a practitioner must go out on a limb to consider that the shot may have had something to do with the reaction. In a lot of cases, these were the same practitioners who recommended these shots in the first place! I'm quite sure Caitlin's cousin's cardiologist didn't send his information to VAERS.

Most practitioners wouldn't report an adverse event unless the event happened within a day or two of the vaccination. However, most incidents we observed were caught due to an incidental finding that occurred weeks, months, or even years later! In other words, a patient will be seen for something unrelated to Covid, and the practitioner will discover something unique. One of my nurses works at our local area hospital. She had an incident at work a few years ago that required her to have an electrocardiogram (EKG). Her EKG came back perfect, with no abnormalities. Later, she was also essentially forced to get two Pfizer shots to retain her job at the hospital and wasn't very happy about it. Six months after receiving both shots, she needed a surgical clearance for a hysterectomy. She had no symptoms and no pain, but the doctor called and told her that her EKG looked like she had a myocardial infarction.

The doctor would not clear her for surgery unless she saw a cardiologist. When she went to visit the cardiologist, he did an echocardiogram of her heart. This showed that she had right atrial hypertrophy (enlarged heart), specifically, the right atrium. The first thing he said to her was, "When did you get the shot?" The cardiologist told her that he'd been seeing a lot of anomalies on incidental findings since the vaccines came out. He cleared her for surgery, and the hysterectomy went well. Recently, though, she has experienced some additional small issues. She had a CT scan for an unrelated incident, and that same cardiologist called her to tell her that she had developed plaque in her aorta, which was also an incidental finding. They were not looking for it, but he noticed it. To top it off, he compared this CT scan to another from 2020, where he discovered this plaque was not previously there.

I had dinner with this same cardiologist a few months ago, and we had an interesting conversation. During the dinner, he said something to me, and I reacted, maybe a little rudely, but I wanted to make a point. We were talking about the vaccines, and he said to me, "I used to recommend them all the time, but not anymore. Once I started seeing issues with them, I stopped."

I looked right at him and said, "Are you fucking kidding me?" His head snapped up, and he couldn't believe I was challenging him, but he quickly wanted to know where I was going with this. I told him, "You are a cardiologist; we are primary care. When we have an issue outside the scope of what we do, we refer them to others, such as you, for your expertise. We have tested over nineteen thousand patients, done over forty thousand tests, and do work with Covid patients all day, every day. Maybe next time you don't know the answer to the question, you should say just that. Tell them, 'I don't know,' and refer the patient to someone who does. You will lose credibility with your patients, and it costs all of us in the long run."

I didn't know what his reaction was going to be, and I'll admit, I was a little harsh. He looked right at me and said, "You're absolutely right; I just didn't know, and my patients were looking for answers. I wish now that I had just told them I didn't have enough data, but I was following the CDC guidelines."

All manner of strange things has happened to patients since the beginning of mass vaccinations. One of our patients was in her early forties and was at the pinnacle of health. She was an avid runner, a busy mom of five kids, and worked out every morning without fail. She had booked a trip months before COVID-19 started and felt forced to get the vaccinations to be able to travel and go on the trip with her family. She and her entire family were adamantly against getting the vaccinations, and she had not previously had Covid as far as she knew. A week after receiving the vaccine, she woke up and could not put weight on her right leg. She was experiencing pain, her leg was warm to the touch, and she felt like she had a Charlie horse that wouldn't go away. She was rushed to the emergency room and was diagnosed with a Deep Vein Thrombosis—a clot.

Her clot was so severe that it had caused death to the surrounding tissue and muscle in her calf. After a week in the hospital and several failed treatments, the surgical team performed a fasciotomy on her calf. This is the process of removing muscle and tissue from the area to relieve swelling and pressure, and the patient lost a quarter of her calf and left the hospital with a wound vac. She needed extensive rehabilitation and physical therapy, and a second surgery for closure. She had asked if anything could be done regarding the shots and asked whether this would be reported because, in her opinion, her clot was directly related to her vaccination. She also asked if there was anyone she could bring legal action against, but the government, the pharmaceutical companies, and the practitioners who administer these vaccines are all immune from prosecution.

We hear stories like these every day, but none of them really hit home until it's someone you love, like in the case of Caitlin's thirteen-year-old cousin. In my case, it was my dad. I've seen too much cause and correlation over the last three years to ever recommend a vaccine. First, this virus is 99.98 percent survivable without any medical intervention in the first place. My team and

I have never lost a patient following vaccinations, but I've seen enough issues such as those I discussed here to warrant me steering patients in the opposite direction.

The only instance I might recommend a shot is if you are at a point in your life where a cold or flu could kill you. However, I would recommend you get one ONLY before the peak seasonality of Covid in your area, and the ONLY vaccine I would recommend out of the main three would be J&J. Nevertheless, I would caution those patients that based on our data, the vaccine could end up causing more damage, and you may be better off without it. When you get a chance, go look at the VAERS data on adverse events. Check out the numbers and realize that what you see is only a microcosm of what made it to that reported level.

Remember when I said it was pure malpractice to tell patients that they wouldn't be "as sick" from COVID-19 if they were fully vaccinated? This was after the government and the CDC told you that you couldn't catch it if you "were fully vaccinated." When we look at the pre-vaccine data, we see that 85 to 90 percent of our positive patients were asymptomatic or mild. Post-vaccine, the numbers are EXACTLY the same. The data doesn't lie! While the number of patients who died early on during the pandemic were unvaccinated. As of this writing, most of the patients dying in the hospital are fully vaccinated people. What happened to the term "pandemic of the unvaccinated"?

CHAPTER 13

My Dad and Common Sense

My dad was a relatively healthy individual. For the record, he had a stint put in one of his coronary arteries many years ago when he was in his forties, but overall, he was very healthy. He was a frequent visitor to the doctor simply because he kept up with his health. My dad was an avid chef who loved to cook. He was active and well taken care of. He stopped working in his fifties and never had to work once he retired. He enjoyed a very long career with the Department of Defense and was among the top brass on a civilian level. He smoked many years ago but had been smoke-free for the better part of twenty-five years. He had been given a clean bill of health by his cardiologist just two years ago as of writing this book, and when he moved out of Fort Myers, Florida, up to Gainesville, he promptly found a reputable internal medicine doctor as well as a new, highly recommended cardiologist in the

Gainesville area. My dad kept up with his health like it was his job, besides making some of the best food on the planet.

I loved my dad but never got to see him as often as I'd have liked to. We were busy, and sometimes the distance was a little challenging, but we did talk every week or two. Over the course of the pandemic, I saw him once or twice, but the virus and the circumstances surrounding it put a strain on visitations. I often told my father about what we were seeing in the office, but like any parent, he took what his son said with a grain of salt. I said to my dad on multiple occasions, "Dad, you do not want this shot; you do not need this shot. If you saw what I see in my office every day, you would run far away from this thing." He laughed it off because what I was saying wasn't what he was hearing every night on MSNBC.

"That's not what everyone is saying, Mike," he would say.

I told him, "Dad, there's a lot of us saying exactly what I'm saying, but the news isn't picking it up."

My father passed on the day before Valentine's Day, February 13, 2023. When I got the call from my stepmother, she was in shock. The autopsy report returned a couple of weeks later, and it said he had an enlarged heart, twice the size it should have been, a clot in his coronary artery, and his kidneys looked like they were about to fail. Coincidentally, my sister, who also had a series of vaccine shots, is experiencing the same heart and kidney issues. As of writing, she is due back at her nephrologist next week. My dad was a good man who was influenced not by data but by rhetoric and conjecture. He was a smart man; however, he put his faith in the government's narrative. I fully believe, after everything I've seen, that his vaccinations were linked to his death.

I said something to my immunologist a few years ago, right when we first started seeing issues stemming from the vaccines and the "breakthrough" cases came out. I told him I had a hypothesis that if someone never had Covid but got a shot, had a small inflammatory reaction from that shot, then got another shot and had the same reaction, then got a booster shot and experienced another inflammatory reaction before contracting Covid, would the patient theoretically get sicker from the—He stopped me right there and said, "from the cytokine storm" and told me it was a great hypothesis. He'd taken the words right out of my mouth! For the record, my immunologist was adamantly in favor of getting his vaccine shots but didn't analyze data as we do or look at cause and correlation. His job is to analyze samples, so all he sees every day are Covid-positive cases. He was and is a great resource, as I would speak with him weekly to discuss trends and learn more than I thought I would ever know about immunology. On that very same call, he told me I should be an immunologist, to which I replied, "No thanks; that's why I have you!"

The side effects, conditions, and complications we're seeing today as a result of the vaccinations reflect the hypothesis I had in that discussion with my immunologist. While the CDC and White House kept saying that "this is a pandemic of the unvaccinated," I knew that was the farthest thing from the truth. I told my staff that the data from the vaccinations just need to catch up, and you will hear the opposite of that statement soon.

Now, most patients who succumb to COVID-19 are FULLY vaccinated, and from what I've seen, their family members are beginning to rethink what they've been fed all these years. My team and I were telling our patients this right from the beginning! This information was out there, and if I'd had some backup from the media and government, my dad might still be alive today.

My dads' case will never make it to VAERS because there is no way to prove that his death was from the vaccine, and neither will the vast majority of cases such as his. Most people will be fine after receiving the vaccines; they will have no complications or adverse reactions. I say that in a general sense as we do not have any long-term data to show how those vaccines will affect patients over the course of years. However, the same can be said for most people who contract COVID-19—they will be fine and won't experience any serious complications. The point then becomes: why would people put themselves in a position to receive multiple reactions when they don't need to? It astonishes me when a patient contracts Covid and then rushes out to get a vaccine a month later. You're good; you don't need it! You've got natural immunity. Go take that cruise and relax.

The Covid vaccines have been proven to work like flu shots in effectiveness and duration. I commonly ask my patients whether they would get four flu shots in one year. Think about this. Covid causes inflammation which is bad enough. The shots also cause inflammation. If you ran out to get four Covid shots in one year and then still caught Covid afterward, it's like getting it five times. I'm sure your body wouldn't be too happy if you had the flu five times in one year! Why are you putting an experimental vaccine into your body that can cause you more harm for something that is completely survivable? It doesn't make sense; not to me, and not to my team. It directly conflicts with our data, which has been SPOT ON since early 2020.

I play a lot of baseball and also work with a lot of former players. I was having a conversation with a former pitcher in November 2021 who was one of my favorite players back in the day. He is one of the nicest guys on the planet and is a true gentleman. We got onto the topic of COVID-19, and he started picking my brain. When I saw the look on his face after answering some of his questions, I could tell what channels he was watching for his news sources. He listened respectfully to all my answers, but at the end of our conversation, he said, "Mike, why are you the only one saying all this?"

I said, "I'm not! Far from it. There are plenty of us saying it, but the news won't report it." We ended our conversation there, but it reminds me of the revelations that came out on Twitter in 2023. It is clear now from multiple independent reporters (most of who were also touting the opposite of the media's narrative) that all of us who had a different narrative were being censored, shadow-banned, and silenced. Those news organizations, social media platforms, and "experts" who had no knowledge of what they were talking about are culpable for the mess we have seen.

The media was confusing everyone. CNN had their death count right at the top of the screen because they would rather focus on sensationalism instead of disseminating helpful information, at least in my opinion. That network was completely wrong on almost everything, and they would televise Governor Andrew Cuomo's daily briefings—a guy who put Covid-positive patients back in nursing homes! That was a brilliant move if you are TRYING to kill people. One of our nursing homes, the one that had us come in early for testing, did very well during the pandemic as far as managing positives and keeping people safe, though they had an initial fifteen deaths in their patient population before we started working there. However, they got their patients all the way to the point of getting vaccinated without further incent until . . . immediately after getting all their patients vaccinated, six of them experienced clots, and three died from stroke complications within the two weeks that followed.

I'm not anti-vax. In fact, I had a Hep B booster right before the pandemic because I work in the medical field and wanted to make sure I was protected. That booster is an attenuated vaccine that has been proven in use for many years. Yes, some recipients have side effects, but the benefits outweigh the risk. What I have seen with the Covid vaccines, especially the mRNA vaccines, is completely on the opposite spectrum. I could tell you hundreds of stories to emphasize my point, but those stories are purely anecdotal. It's the *data* that drives the point home. My early hypothesis was that by mass vaccinating the entire world with these unproven and ineffective vaccines, you could estimate that we have shortened the lifespan of the entire populus by about three to five years on average, and that's a conservative estimate. I would garner to say that it could be as high as seven to nine years.

If you take just one thing away from this book, remember this: medicine is NOT one size fits all. The CDC was recommending the same dosage for a fifty-pound kid as it was for a two-hundred-and-fifty-pound man. They forced this on kids before allowing them to go to school. They forced this on anyone, regardless of clotting factor. If I had a patient with a chance of clotting, they might not be a good candidate for the vaccine due to the risks. Nobody was conducting D-Dimer tests (a test to check for blood clotting problems) before a questionable patient was administered a vaccine. At the time of writing, the CDC has withdrawn its recommendations for kids to get the vaccine just this week. They said that the risk is not worth the reward. Tell that to Caitlin's cousin and his family.

Caitlin has told that story to lots of patients to try and open their eyes, but it's hard to fight City Hall, the White House, and the whole of the media at the same time. The government and the media will eventually catch up, right? They are starting to, but it's too late for some. The mass vaccination of the entire population without data, tests, or any semblance of logic is the most RIDICULOUS thing I have ever seen in my career. The masks were a distant second, and that's because the masks won't kill you. They just make you look stupid. The vaccines, though, didn't save lives. It is a COMPLETE fallacy to think that you won't get "as sick" as someone who didn't get the shot once you are vaccinated. Remember, the vaccines only give you a spike protein response and NOT to the entirety of the virus. You will NOT get those B and T cell responses from the vaccine alone. If you keep giving yourself inflammatory reactions, you can cause permanent damage.

The reaction someone gets from the virus and the shot sincerely sucks, even if it's not full-blown sickness. Some patients get earaches every time they have Covid; some develop heart issues, and some get joint issues. We find that cytokines rush around your system for at least two weeks. Cytokines attack areas of instability and weaken already weak areas of your body. While you may not have developed joint, heart, or complications to date, it may be something that can develop later in life. Now those cytokines are causing damage to those areas that may be susceptible. Any underlying issue that may not develop until you are older may now start to propagate at a much younger age.

If you observe the damage in an autopsy report from a patient who has passed right after receiving the vaccine, you will see things that are not supposed to be present. Pathology reports show massive areas of inflammation in the liver, heart, lungs, and kidneys. This is because those cytokines are rushing around and causing damage in areas that may already be weak. The mitochondria of the cells become fragmented and damaged when they normally look smooth and organized. The mitochondria are what give your cells energy to function. When you damage the mitochondria, you can feel weak and run down. That's why we decided to try NAD for cellular repair. It's a short-term fix to a larger problem.

Some ask whether issues like pericarditis and myocarditis will resolve themselves with time. We simply don't know the answer to that yet as we don't have enough long-term data, but I bet you wish you could avoid those issues. I'm sure most of you have already had COVID-19 at this point, and most of you are fine. However, some people develop long Covid, where some symptoms linger. I have a friend who lost her sense of taste and smell and never got it back. Covid has permanently damaged her olfactory nerves. There are patients with permanent heart, joint, and lung issues, too. I never said that COVID-19 isn't real or dangerous. I'm sure some of you even know someone who lost their life due to the pandemic. I get it; I've been through it all with you every step of the way. My point relating to vaccines and common sense is that the data speaks for itself, and we should all take the time to look at the totality of the circumstances. Step back and look at that thirty-thousand-foot view. Some people won't see the forest for the trees, but when you back up and look at the entire picture, it all starts to come into focus.

I remember driving to work one morning and listening to the news on my radio. A prominent physician who is a contributor to one of the national networks I like was misinforming the masses. I wanted to figuratively jump into my radio just to make her stop talking. She was talking about the vaccines and, in particular, "breakthrough cases." We never used that term in our office because we understood you could still contract Covid while being fully vaccinated. She was trying to implore people to get vaccinated because, in her words, "breakthrough cases were extremely rare." She said, "*There are only five thousand cases of breakthrough cases in the entire country, and the benefits clearly outweigh the risks."* She was trying to make the point that the vaccinations worked. We share data with other practitioners in our area all the time, and just the day before I heard this, we had five breakthrough cases in my office alone, and the urgent care had three. Those were just eight that I knew of out of only two clinics. We knew that this was a rampant issue already.

Her blatant misrepresentation of what was happening on the street was part of the problem. She didn't say the words "I don't know." Instead, she tried to appear intelligent to earn her keep. I had a lot of issues with most of the doctors on television. There were prominent doctors I had put a lot of faith in over the years, which had now lost credibility in my eyes. I won't mention their names, but I will mention the one doctor who was usually on point. The one guy who was on point during the entire pandemic was Dr. Marty Makary. For the record, I do not know him and have no interest in mentioning his name. I just like to give credit where credit is due. He was the only one I could point my patients to with a high degree of comfort. If my data wasn't front and center, at least there would be someone in public speaking with some common sense. Good job, Marty!

There were other practitioners in the industry who were speaking very loudly about what they were seeing, but major networks didn't put them on air. Some of these practitioners and researchers were "canceled" or discredited entirely by the mainstream media. If anyone is culpable for the death, destruction, shutting down of businesses, and disruption this pandemic caused, it is the media. In my day, the media questioned things; they challenged the status quo and forced a conversation with healthy debate. Today's media seems complacent and cooperative with whatever the by-line of the federal government is. It's a shame because the truth would have saved a lot of time, energy, and lives. There seemed to be a lack of common sense throughout the entire pandemic and, three years later, the truth only changes the narrative once it becomes blatantly obvious and the masses have no choice but make it mainstream. What a shame, what a farce, what a disaster. We could have solved this problem in a matter of months, not years!

My dad, Caitlin's cousin, my buddy Mike's dad—none of them can get their lives back. My patient with her calf removed can't run as she used to, and all my patients with serious medical issues can't get their health back. I bet most of you already know a lot of what I'm telling you now that you have had some time to digest Covid for yourself. I bet if I told you all of this in 2020, 2021, or even 2022, you would have questioned some of it. However, all we can do is move forward as a society, learn from our mistakes, and hope to God nothing like this response to a pandemic never happens ever again . . . but it will, if human nature tells you anything about groupthink and how society reacts to pressure. Common sense is hard to come by!

CHAPTER 14

What Did We Learn and Where Do We Go From Here?

In July 2020, four months into the pandemic, I took a vacation with some friends. While I was there, I got a call from a good friend of mine who is very intelligent. He told me that he was having some pain in his leg, and it had a red spot that was warm to the touch. He asked me, "Do you think it's Covid?"

I laughed and said, "No, but let me put you on the phone with one of my practitioners." My practitioner spoke with him and promptly sent him off to the hospital. They evaluated him, diagnosed him with cellulitis, and kept him overnight. They treated him with IV antibiotics and released him. Boring story, right? But it highlights the pure nonsense we dealt with in the early stages of the pandemic. Everybody thought everything was Covid!

I just had dinner this week with a group of friends. Every month, six of us go out to a popular restaurant in New Jersey. I was talking to my friend's wife about Covid, and she told me that her husband had caught it and was sick for a few days. Keep in mind that he is fully vaccinated. She told me that she NEVER had it. I said, "How do you know?" She told me that she never had a symptom and that she tested negative on a rapid. I told her that in most cases if you don't have any symptoms, there would not be enough virus in her system to warrant a reaction on a home rapid test. I also told her that 90 percent of my over four thousand positive patients were completely asymptomatic. I then told her that if someone was symptomatic in a household, our studies showed that 100 percent of the people in the household also tested POSITIVE. The only way for her to truly know would have been to test every week in that timeframe with a PCR test. She disagreed and insisted her home test was accurate and that because she never had a symptom, she never had Covid.

I have this conversation every day. Patients don't have any experience with Covid, have no data on it, and don't understand the totality of the circumstances, but they are very insistent on what they think they know. We are at a point now where people don't care about the truth, and they don't care about the data. However, if the talking heads on TV said we were having a resurgence, they would mask up and stay at home. These same people would rush out for their ninetieth booster and social distance from the world. We've spent a lot of time becoming experts on COVID-19 to protect those who don't have the same data that my team has access to. Those same lay people are very adamant in their "beliefs" regarding

Covid, which is what usually controls the narrative. It's pure symbolism over substance.

Remember when someone sneezed in an airport, and everybody would turn around to stare at the offender as if to say, *why are you trying to kill us, jerk?* The pandemic turned a nation of good people into a bunch of scared, mind-numbed, paranoid terminators. It was neighbor against neighbor out in the trenches. If you were standing too close to someone, you might get yelled at for encroaching on the six-foot zone. If your mask fell under your nose when you fell asleep on a flight, you would be woken up by a team of flight attendants. Some people were even arrested for surfing in the OCEAN! Is this really what our society turns into when chaos befalls us?

This entire thing was so sad, in my opinion, that I hope I never have to see another pandemic in my lifetime. Are we really living among a population that makes assumptions, follows orders (no matter how ridiculous), and suspends science and reality? We could have saved a LOT of time and lives if the masses heard my data in April 2020. My team and I only needed one good month of data to determine that COVID-19 wasn't as serious as everyone thought it was. Sure, the treatment regimens needed time to be worked out, but once we knew that the masses wouldn't succumb to this sickness, we should have hit the brakes on all the nonsense. If you understand the timeline of a virus, which you do now, you know that if you stopped everyone in the world from moving about for about twenty-one days, you could theoretically stomp out the spread of Covid. However, that's not what people were doing, and I laughed at those who posted "Stay home, it saves lives" on their profiles.

I also found it funny when people posted pictures with their masks on. For a society so engaged in self-gratification, people who need to put posts on social media just for some attention, it was comical to see all the posts of people in "quarantine." Yeah, we get it, jerk off; we're all in the same boat! I get that people were scared, and I don't blame them based on the information they were receiving; however, they didn't need to be as nervous as they were. A little education goes a long way, especially when the entire world is this frightened. Once people started contracting Covid, got over it, and recovered from their mild symptoms, I think the masses started to catch onto what we were talking about. Why, then, did the ridiculousness persist?

I walked into an orthopedic office for a second opinion on a scheduled knee surgery during the pandemic. I honestly forgot my "mask" in my car, and I was scolded upon entry. They had one of those plastic screens in between the staff and patients—the kind that you could talk around or through a hole in the center. After being reprimanded and observing the flimsy measures they had set up to placate their patients, I had had enough. I said, "Hand me back my paperwork!"

The office manager looked at me and said, "Excuse me, sir?"

I said, "I'm leaving. I came here for medical advice, and it's obvious that you don't understand science, so now I'm questioning your medical prowess!"

As I mentioned earlier in the book, I said the same thing to my attorney, who had one of these plastic barriers that were about two feet by two feet sitting on the receptionist's desk. It was time we called out stupid for what it is! *Thanks for trying to make me "feel" better, you're just insulting my intelligence, and now, I have serious doubts about yours!* I still see it every day in airports, public places, and private offices. Just today in March of 2023, one of my practitioners told me that in her other urgent care office, they finally stopped requiring the staff to wear masks. To which I replied, "Took them three years. I feel bad for their patients."

I can tell the masses really don't care about COVID-19 anymore because no one comes into our office for it. It didn't tail off slowly or noticeably decline; people simply stopped coming. The number of Covid cases at the end of 2021 were staggering and probably the highest we saw during the pandemic. After that surge came and went, we had a slowdown but returned to our normal volume of steady patients. Then, suddenly, in the middle of 2022, those numbers went down to zero. Covid became what I had predicted it to be, another respiratory pathogen buried in a panel that nobody cares about.

The fascination and sheer panic with Covid have turned into a malaise and frustration for people now. Nobody wants to talk about it anymore, and most people have some level of PTSD lingering from the lockdowns, stress, and thought of years wasted while we gave the powers that be the time to fix the issue. I feel for those families who couldn't see their loved ones because they were too scared to. I saw way too many birthday parties from afar while I sat at one of our nursing homes, and their families had to wave through windows. I also witnessed a lot of those nursing home patients die alone, not of the virus, but due to restrictions. I have too many memories of long conversations with patients who didn't have the resolve or scientific knowledge that I had. Those patients sincerely thought they were doomed and stopped living their lives.

I had the same conversations with patients night after night. I wish I'd had time to write this book earlier so I could have just handed it to patients to ease their fears. Many thought I had lost my marbles at the time, but once you sit with people, explain the science, and paint a picture, most patients get it and feel a lot more confident. I lost faith in the CDC, as most of you probably have. I lost faith in our medical community, as I'm also sure most of you have. I commend the doctors and nurses who went into the unknown and really made a difference, but I question the authenticity of those who closed their offices the moment their patients needed them most during the pandemic. I am baffled by the practitioners who sent patients to a testing center so they could show a negative test before being seen by their own primary care doctor! Way to come through for your patients.

This is all without mentioning the long-term effects that those office closures had on their patients. People started neglecting their health dramatically during the pandemic. There is more heart disease now than ever before. There are more diabetic amputations now than ever before because patients neglected their wounds. More people have cancer that has spread too far because those patients weren't being checked. There are more hospice patients now than ever before. Instances of hypertension, kidney issues and patients on dialysis, obesity, and mental health issues have all grown to unprecedented levels. I could go on and on. We did much more damage with the response to COVID-19 than the virus itself did. Quite frankly, we've learned that our society did a really shitty job managing a pandemic!

I remember President Trump saying very early on that he didn't want the cure to be more deadly than the virus. He was 100 percent correct in his assessment, and it's too bad the people surrounding him didn't heed his warning. Sometimes, instead of reacting, we need to calmly sit back, think about the totality of the circumstances, and make sound judgments. Once election season started, those sound rational arguments went out the window for a more partisan approach. I'm not saying that to spur a political argument. What I'm saying is the arguments made for more masks, more lockdowns, and arresting people in parks and on beaches made ZERO sense scientifically. The arguments did not match the science or the data. Their arguments directly conflicted with the data we had.

There were also a lot of conspiracy theories floating around, like, "Covid killed the flu." Yes, it's true we didn't see many instances of the flu during the first round of COVID-19. However, that's because it was a mild flu season. People then wondered whether the masks played a part in these lowered flu cases. Nope, flu particles are about the same size as COVID-19 and spread the exact same way. It's most likely that having kids out of school for all those months inhibited a bad flu season, but it really doesn't matter. We saw the flu return the next season and *still* had COVID-19 to deal with.

Having been surrounded by the data and seeing patients with Covid, I also became great at predicting how patients would react. Remember my friends Joe and Carnella in Key West? I got a call from Joe one day who told me his entire house was sick: him, his wife, and his son. I went to his house and tested all three of them. His son was very nauseous, so we helped him with some regimens and recommended a bland diet. Joe is a big guy; his wife is very slim and in shape, and his son is about nineteen years old. They all got sick at the same time, and yes, they all came back positive on PCR tests. I looked at them and said to his son, "You will feel fine by tomorrow; Carnella, probably the next day, and Joe, you're going to feel like dog shit for a few days, but you'll be OK." I called Joe the next day to ask how everybody was doing, and he told me that his son was out running and felt fine once he woke up that morning. His wife was just a little run down, and he was considering monoclonal antibodies. Overall, Carnella was fine in a day or two, and Joe had a tougher time with his bout but recovered just fine after about a week.

My team and I got good at giving people an expected Covid timeline that was in conjunction with any co-infections, comorbidity, and their overall physical health. It gave people some solace once they understood the bigger picture, and I am proud to say that our office really helped people during the pandemic. I'm also proud of my staff for weathering the storm. They couldn't have performed any better, and I'll be forever indebted to them for acting professionally and dealing with all the nonsense that came along with the situation. There are only so many times you can get screamed at for not wearing a mask or for a FedEx or UPS delivery being delayed. When you do nothing but handle Covid all day, every day, it can start to wear on you.

So, where do we go from here? First, I would say that the next time you take advice from an "expert" throwing suggestions at you, ask what evidence their advice is based on. You should ask them if they have any experience with the matter at hand, whether they have any data, and consider if they are just regurgitating talking points from another unreliable source. We've seen enough "experts" on TV to realize that their degrees aren't worth the paper they're printed on. These "experts" have all changed their stories now, but none of us have three years of our lives to let those idiots sort it out. Stop relying on them so much and rely on yourself!

Second, try to get both sides of the story. There were two sides, even three or four sides to the pandemic—but Dr. Fauci was not considering the other real experts opinions on Covid-19 for some reason. He didn't want to consider the data that was coming in from offices like mine. His sitting in Nationals Park with two or three masks on while he was OUTSIDE was comical at best. It insulted all our intelligence, but it swayed the most vulnerable of people. When I see people in the airport now with an N-95 mask falling off their faces, I roll my eyes, and one word comes to mind: MORON! Did Dr. Fauci create the moron in that person, or did it just bring it to light? It was probably a combination of the two. When I see people rushing into practitioners' offices to get their seven-year-old vaccinated, I feel bad for the parent, but I *really* feel bad for that kid. As I mentioned earlier, the CDC just changed its opinion on vaccinations for kids. They say they don't need them. I would tell you; kids don't need them and add that they should run far away from them.

You see, my staff and I are the ones who get burned at the stake because the mob has been taken over by groupthink. It is a good, sound debate that spreads ideas and dialogue. Silencing that dialogue doesn't solve any problems; it just creates more of them in the long run. We've got a long way to go because I don't think that as we are now, we've learned a single thing from this pandemic, even with the masses coming around to the facts. I think if the next pandemic came knocking at our door right now, we would be doomed, and mostly by ourselves. You should be angry that we knew *what* we did *when* we did, and the media and government didn't let you hear it. You should demand accountability from your government and the fourth establishment, the media.

If you were vaccinated and aren't feeling yourself, please go to your practitioner and get a checkup. Get that EKG you've been holding off on, and get your bloodwork checked. Most people, as I mentioned, won't be affected adversely by those vaccines, but if you were, it might cause you a health concern. Covid sucks, and I don't want to get it again, but I'm bound to get it multiple times throughout my lifetime. I've only had it three times that I know of. However, I refuse to stop living my life, eating at my favorite restaurants, and traveling to some great destinations for fear that being around people may infect me. Life is too short to stop living because of fear.

If I could do one thing differently, I wish I could have finished this book earlier so my dad could have read it. It may have saved his life! In his memory, and in the memories of everyone who passed during this pandemic, I hope this book can save your life or the life of someone you love.

ACKNOWLEDGEMENTS

First and foremost, I would like to acknowledge all our patients for without you, there would be no story to tell! Kelly, thank you for always having my back and keeping me on track to get this project done! Vincent, you are a good boy and great company when I write on the couch! Caitlin for running a tight ship and keeping my life in order during all the nonsense. To my entire staff past and present, you are all valued more than you know. Covid was a unique and trying time in all our lives and it will always connect us. You stuck your neck out and weathered the storm without knowing what you were getting into at the time. Great job and thank you from the bottom of my heart! To Paul Rotella, great catches and I am glad you are on the mend buddy! Carnella, I had no idea how hard your job is and what goes into publishing a book! Thank you for the guidance and direction. To all my friends and family and patients who encouraged me to write this book, I hope it helps you put it all into perspective. And lastly, to my dad, words can't describe how much I miss you. You gave me the strength and resolve to finish this project and I will always be grateful to be your son!

ABOUT THE AUTHOR

Dr. Michael J Schwartz has been an entrepreneur since 1993. Over the course of his career, he has owned and operated many types of very diverse companies. He is an accomplished private pilot and an avid New York Yankee fan. A former police officer, he developed and teaches a course, "The Secrets of Body Language and Communication" to private and governmental entities around the world. He has been performing stand-up comedy since his early 20's and performs regularly. He has appeared on multiple television and radio shows. Schwartz founded the charity Hometown Heroes in 2008, which is credited with distributing over three million in funds to those less fortunate. Hometown Heroes has been recognized by multiple entities and was even awarded the prestigious Robin Hood Service Award. He has sat on numerous boards and has won many distinguished awards for his business career and philanthropic work. He is a recipient of a Paul Harris Fellow from Rotary International. He was recognized by the United Way and received their "Top 40 under 40" award. As a police officer he was decorated with both a "Class A" and "Class C" meritorious service award. He resides in Wall, New Jersey, and Tampa Florida.

Made in the USA
Middletown, DE
25 September 2023

39025506R00102